Pulse-width Modulated DC–DC Power Converters

Solutions Manual For

Pulse-width Modulated DC–DC Power Converters

MARIAN K. KAZIMIERCZUK
Wright State University,
Dayton, Ohio, USA

A John Wiley and Sons, Ltd, Publication

Other Wiley Editorial Offices

John Wiley & Sons Inc., 111 River Street, Hoboken, NJ 07030, USA

Jossey-Bass, 989 Market Street, San Francisco, CA 94103-1741, USA

Wiley-VCH Verlag GmbH, Boschstr. 12, D-69469 Weinheim, Germany

John Wiley & Sons Australia Ltd, 42 McDougall Street, Milton, Queensland 4064, Australia

John Wiley & Sons (Asia) Pte Ltd, 2 Clementi Loop #02-01, Jin Xing Distripark, Singapore 129809

John Wiley & Sons Canada Ltd, 6045 Freemont Blvd, Mississauga, ONT, L5R 4J3

Wiley also publishes its books in a variety of electronic formats. Some content that appears
in print may not be available in electronic books.

British Library Cataloguing in Publication Data
A catalogue record for this book is available from the British Library

ISBN 978-0-470-74101-6

Typeset by Laserwords Private Limited, Chennai, India

Contents

Chapter 1
Introduction

.1 A voltage regulator experiences a 100 mV change in the output voltage, when its input voltage changes by 10 V at $I_O = 0.2$ A and $T_A = 25\,^\circ$C. Determine the line regulation and the percentage line regulation. The nominal output voltage is $V_{Onom} = 3.3$ V.

The line regulation is

$$LNR = \frac{\Delta V_O}{\Delta V_I} = \frac{100\,\text{mV}}{10\,\text{V}} = 10\,\frac{\text{mV}}{\text{V}}.$$

The percentage line regulation is

$$PLNR = \frac{\dfrac{\Delta V_O}{V_{Onom}} \times 100\,\%}{\Delta V_I} = \frac{\dfrac{100 \times 10^{-3}}{3.3} \times 100\,\%}{10} = 0.303\,\%.$$

.2 A voltage regulator is rated for an output current $I_O = 0$ to 50 mV. Under no-load conditions, the output voltage is 5 V. Under full-load conditions, the output voltage 4.99 V. Find the load regulation, the percentage load regulation, the dc output resistance, and load/line regulation.

The load regulation is

$$LOR = \frac{V_{O(NL)} - V_{O(FL)}}{\Delta I_O} = \frac{(5 - 4.99)\,\text{V}}{50\,\text{mA}} = 0.2\,\frac{\text{mV}}{\text{mA}}.$$

Hence, the dc output resistance is

$$R_o = \frac{\Delta V_O}{\Delta I_O} = LOR = 0.2\,\Omega.$$

The percentage load regulation is

$$PLOR = \frac{V_{O(NL)} - V_{O(FL)}}{V_{O(FL)}} \times 100\,\% = \frac{5 - 4.99}{4.99} \times 100\,\% = 0.2\,\%.$$

Pulse-width Modulated DC–DC Converters – Solutions Manual   Marian K. Kazimierczuk
© 2008 John Wiley & Sons, Ltd

The line/load regulation is

$$LLR = \frac{\frac{\Delta V_O}{V_{Onom}} \times 100\%}{\Delta I_O} = \frac{\frac{0.01}{5} \times 100\%}{50\,\text{mA}} = \frac{4\%}{\text{A}}.$$

1.3 A series linear voltage regulator is operated under the following conditions: $V_I = 6$ t 15 V, $V_O = 3.3$ V, and $I_O = 0$ to 0.4 A. Find the minimum and maximum efficiency c the voltage regulator at full load.

The minimum efficiency of the series voltage regulator at full load is

$$\eta_{FL(min)} = \frac{V_O}{V_{Imax}} = \frac{3.3}{15} = 22\%.$$

The maximum efficiency of the series voltage regulator at full load is

$$\eta_{FL(max)} = \frac{V_O}{V_{Imin}} = \frac{3.3}{6} = 55\%.$$

1.4 A voltage regulator has $R_L = 10\,\Omega$, $V_I = 10$ V, $V_O = 5$ V, and $\eta = 90\%$. Find the d input resistance.

The dc voltage transfer function is

$$M_{V\,DC} = \frac{V_O}{V_I} = \frac{5}{10} = 0.5.$$

The dc input resistance is

$$R_{in(DC)} = \frac{\eta R_L}{M_{V\,DC}^2} = \frac{0.9 \times 10}{0.5^2} = 36\,\Omega.$$

Chapter 2
Buck PWM DC–DC Converter

2.1 Derive an expression for the dc voltage transfer function of the lossless buck converter operating in CCM using the diode voltage waveform.

The voltage at the input of the output low-pass filter L–C–R_L is $v_{AB} = -v_D$. Thus,

$$v_{AB} = V_I, \quad \text{for } 0 < t \le DT,$$

and

$$v_{AB} = 0, \quad \text{for } DT < t \le T.$$

The average value of the voltage across the inductor in steady state is zero. Hence, the dc component of the output voltage is equal to the average voltage at the input of the output filter:

$$V_O = \frac{1}{T} \int_0^T v_{AB}\, dt = \frac{1}{T} \int_0^{DT} V_I\, dt = DV_I.$$

This gives

$$M_{V\,DC} = \frac{V_O}{V_I} = D.$$

2.2 A buck converter has $V_I = 22$ to $32\,\text{V}$, $V_O = 14\,\text{V}$, $I_O = 0.2$ to $2\,\text{A}$, and $f_s = 40\,\text{kHz}$. Find the minimum inductance L required to maintain the converter operation in the continuous conduction mode.

The maximum load resistance is

$$R_{Lmax} = \frac{V_O}{I_{Omin}} = \frac{14}{0.2} = 70\,\Omega.$$

Pulse-width Modulated DC–DC Converters – Solutions Manual   Marian K. Kazimierczuk
2008 John Wiley & Sons, Ltd

The minimum duty cycle for a lossless buck converter is

$$D_{min} = M_{V\,DCmin} = \frac{V_O}{V_{Imax}} = \frac{14}{32} = 0.4375.$$

The minimum inductance is

$$L_{min} = \frac{R_{Lmax}(1 - D_{min})}{2f_s} = \frac{70 \times (1 - 0.4375)}{2 \times 40 \times 10^3} = 492.1875\,\mu\text{H}.$$

Pick $L = 550\,\mu\text{H}$.

2.3 For the converter given in Problem 2.2, find the transistor and diode voltage and curre stresses.

The voltage stresses are

$$V_{SMmax} = V_{DMmax} = V_{Imax} = 32\,\text{V}.$$

The current stresses are

$$I_{SMmax} = I_{DMmax} = I_{Omax} + \frac{\Delta i_{Lmax}}{2} = I_{Omax} + \frac{V_O(1 - D_{min})}{2f_s L}$$

$$= 2 + \frac{14 \times (1 - 0.4375)}{2 \times 40 \times 10^3 \times 0.55 \times 10^{-3}} = 2.179\,\text{A}.$$

2.4 A buck PWM converter has $V_I = 10$ to $14\,\text{V}$, $V_O = 5\,\text{V}$, $I_O = 0.2$ to $1\,\text{A}$, $f_s = 200\,\text{kH}$ $L = 100\,\mu\text{H}$, $C = 100\,\mu\text{F}$, and $r_C = 20\,\text{m}\Omega$. Find the ripple voltage V_r and (V_r/V_O) 100 %. Also, calculate the ripple voltage across the filter capacitance and the corn frequency of the output filter.

First, we need to know what the mode of operation of the converter is. Assumir $\eta = 100\,\%$, the minimum duty cycle is

$$D_{min} = M_{V\,DCmin} = \frac{V_O}{V_{Imax}} = \frac{5}{14} = 0.3571.$$

The maximum load resistance is

$$R_{Lmax} = \frac{V_O}{I_{Omin}} = \frac{5}{0.2} = 25\,\Omega.$$

The minimum inductance required for CCM is

$$L_{min} = \frac{R_{Lmax}(1 - D_{min})}{2f_s} = \frac{25 \times (1 - 0.3571)}{2 \times 200 \times 10^3} = 40.18\,\mu\text{H}.$$

Thus, $L > L_{min}$ and therefore the converter operates in the continuous conduction mod

The maximum peak-to-peak value of the inductor current ripple is

$$\Delta i_{Lmax} = \frac{V_O(1 - D_{min})}{f_s L} = \frac{5 \times (1 - 0.3571)}{200 \times 10^3 \times 100 \times 10^{-6}} = 0.1607\,\text{A}.$$

The maximum duty cycle for a lossless buck converter is

$$D_{max} = M_{V\,DCmax} = \frac{V_O}{V_{Imin}} = \frac{5}{10} = 0.5.$$

The minimum filter capacitance is

$$C_{min} = \max\left\{\frac{D_{max}}{2f_s r_C}, \frac{1 - D_{min}}{2f_s r_C}\right\} = \max\left\{\frac{0.5}{2f_s r_C}, \frac{1 - 0.3571}{2f_s r_C}\right\}$$

$$= \frac{1 - 0.3571}{2 \times 200 \times 10^3 \times 20 \times 10^{-3}} = 80.362\,\mu F.$$

Since $C > C_{min}$, the ripple voltage is entirely determined by the voltage drop across the ESR. The ripple voltage is

$$V_r = r_C \Delta i_{Lmax} = 20 \times 10^{-3} \times 0.1607 = 3.214\,\text{mV}.$$

Hence, the normalized ripple voltage is

$$\frac{V_r}{V_O} \times 100\% = \frac{3.214 \times 10^{-3}}{5} \times 100\% = 0.064\%.$$

The ripple voltage across the filter capacitance is

$$V_{Cpp} = \frac{\Delta Q}{C} = \frac{\Delta i_{Lmax}}{8f_s C} = \frac{0.1607}{8 \times 0.2 \times 10^6 \times 100 \times 10^{-6}} = 1\,\text{mV}.$$

The corner frequency of the output filter is

$$f_o = \frac{1}{2\pi\sqrt{LC}} = \frac{1}{2\pi\sqrt{100 \times 10^{-6} \times 100 \times 10^{-6}}} = 1.592\,\text{kHz}.$$

2.5 For the converter given in Problem 2.4, the filter capacitance has been reduced to 47 μF. Find the ripple voltage.

The converter is operated in CCM. Since $C < C_{min}$, the ripple voltage depends on the ripple voltage of both the filter capacitance and the ESR. The ripple voltage across the filter capacitance is

$$V_{Cpp} = \frac{\Delta Q}{C} = \frac{\Delta i_{Lmax}}{8f_s C} = \frac{0.1607}{8 \times 0.2 \times 10^6 \times 47 \times 10^{-6}} = 2.137\,\text{mV}.$$

The ripple voltage across the ESR is

$$V_{rC} = r_C \Delta i_{Lmax} = 20 \times 10^{-3} \times 0.1607 = 3.214\,\text{mV}.$$

Hence, the total ripple voltage is

$$V_r \approx V_{Cpp} + V_{rC} = 2.137 + 3.214 = 5.351\,\text{mV}.$$

2.6 A PWM converter operates in CCM at $V_I = 10\,\text{V}$ and $V_O = 5\,\text{V}$. Find the duty cycle D if (a) the converter efficiency $\eta = 100\%$ and (b) the converter efficiency $\eta = 80\%$.

The dc voltage transfer function is

$$M_{V\,DC} = \frac{V_O}{V_I} = \frac{5}{10} = 0.5.$$

For $\eta = 100\,\%$, the duty cycle is

$$D = \frac{M_{V\,DC}}{\eta} = \frac{0.5}{1} = 0.5.$$

For $\eta = 80\,\%$, the duty cycle is

$$D = \frac{M_{V\,DC}}{\eta} = \frac{0.5}{0.8} = 0.625.$$

It can be seen that the duty cycle D increases when the efficiency decreases at a fixed output voltage V_O.

2.7 A buck converter operating in CCM has a MOSFET whose $r_{DS} = 0.025\,\Omega$. The load current is $I_O = 10\,A$. Determine the MOSFET conduction loss at $D = 0.1$ and 0.9.

At $D = 0.1$,

$$P_{rDS} = Dr_{DS}I_O^2 = 0.1 \times 0.025 \times 10^2 = 0.25\,W.$$

At $D = 0.9$,

$$P_{rDS} = Dr_{DS}I_O^2 = 0.9 \times 0.025 \times 10^2 = 2.25\,W.$$

It can be seen that the transistor conduction loss P_{rDS} increases with increasing duty cycle D.

2.8 A buck converter operating in CCM has a diode whose $R_F = 0.025\,\Omega$ and $V_F = 0.3$. The load current is $I_O = 10\,A$. Determine the diode conduction loss at $D = 0.1$ and 0.

At $D = 0.1$,

$$P_{RF} = (1 - D)R_F I_O^2 = (1 - 0.1) \times 0.025 \times 10^2 = 2.25\,W,$$

$$P_{VF} = (1 - D)V_F I_O = (1 - 0.1) \times 0.3 \times 10 = 2.7\,W,$$

$$P_D = P_{RF} + P_{VF} = 2.25 + 2.7 = 4.95\,W.$$

At $D = 0.9$,

$$P_{RF} = (1 - D)R_F I_O^2 = (1 - 0.9) \times 0.025 \times 10^2 = 0.25\,W,$$

$$P_{VF} = (1 - D)V_F I_O = (1 - 0.9) \times 0.3 \times 10 = 0.3\,W,$$

$$P_D = P_{RF} + P_{VF} = 0.25 + 0.3 = 0.55\,W.$$

It can be seen that the diode conduction loss P_D decreases with increasing duty cycle .

2.9 A power MOSFET has $V_B = 0.75$ V, $C_{rss} = 30$ pF, and $C_{oss} = 130$ pF at $V_{DS} = 25$ V. It is used in a buck PWM converter with $V_I = 400$ V and $f_s = 1$ MHz. Find C_{J0}, $C_{ds}(V_I)$, $Q(V_I)$, P_{sw}, $P_{turn-off}$, and $P_{sw(FET)}$.

The output capacitance at $V_{DS} = 25$ V is

$$C_{ds25} = C_{oss} - C_{rss} = 130 - 30 = 100 \, \text{pF}.$$

The zero-bias capacitance is

$$C_{J0} = C_{ds25}\sqrt{1 + \frac{V_{DS}}{V_B}} = 100 \times 10^{-12} \times \sqrt{1 + \frac{25}{0.75}} = 585.95 \, \text{pF}.$$

The output capacitance at $V_I = 400$ V is

$$C_{ds}(V_I) = \frac{C_{J0}}{\sqrt{1 + V_I/V_B}} = \frac{585.95 \times 10^{-12}}{\sqrt{1 + 400/0.75}} = 25.35 \, \text{pF}.$$

The charge stored in C_{ds} at the end of the turn-off transition is

$$Q(V_I) = 2(V_I + V_B)C_{ds}(V_I) = 2 \times (400 + 0.75) \times 25.35 \times 10^{-12} = 20.32 \, \text{nC}.$$

The switching power loss is

$$P_{sw} = 2f_s C_{ds}(V_I)V_I^2 = 2 \times 10^6 \times 25.35 \times 10^{-12} \times 400^2 = 8.112 \, \text{W}.$$

The switching power loss at turn-off is

$$P_{turn-off} = \frac{4}{3}f_s C_{ds}(V_I)V_I^2 = \frac{4}{3} \times 10^6 \times 25.35 \times 10^{-12} \times 400^2 = 5.408 \, \text{W}.$$

The switching power loss in the MOSFET is

$$P_{sw(FET)} = \frac{2}{3}f_s C_{ds}(V_I)V_I^2 = \frac{2}{3} \times 10^6 \times 25.35 \times 10^{-12} \times 400^2 = 2.704 \, \text{W}.$$

2.10 A buck converter has $V_I = 22$ to 32 V, $V_O = 14$ V, $I_O = 0$ to 2 A, and $f_s = 40$ kHz. Find the maximum inductance L required to maintain the converter operation in the discontinuous conduction mode. Assume $\eta = 90\%$.

The minimum load resistance is

$$R_{Lmin} = \frac{V_O}{I_{Omax}} = \frac{14}{2} = 7 \, \Omega.$$

Assuming $\eta = 0.9$, the maximum duty cycle at the boundary between the CCM and DCM modes is

$$D_{max} = \frac{V_O}{\eta V_{Imin}} = \frac{14}{0.9 \times 22} = 0.70707.$$

The maximum inductance is

$$L_{max} = \frac{R_{Lmin}(1 - D_{max})}{2f_s} = \frac{7 \times (1 - 0.70707)}{2 \times 40 \times 10^3} = 25.6 \, \mu\text{H}.$$

Pick $L = 22 \, \mu\text{H}$.

2.11 Design a buck PWM converter to meet the following specifications: $V_I = 12\,V \pm 4$ $V_O = 5\,V$, $I_O = 1$ to $10\,A$, $V_r/V_O \le 1\,\%$, $f_s = 100\,kHz$, $r_{L(dc)} = 50\,m\Omega$, $r_{DS} = 10\,m\Omega$ $C_o = 200\,pF$, $V_F = 0.3\,V$, and $R_F = 20\,m\Omega$.

The converter must operate in CCM because of the load range. The maximum power

$$P_{Omax} = V_O I_{Omax} = 5 \times 10 = 50\,W,$$

and the minimum power is

$$P_{Omin} = V_O I_{Omin} = 5 \times 1 = 5\,W.$$

The minimum load resistance is

$$R_{Lmin} = \frac{V_O}{I_{Omax}} = \frac{5}{10} = 0.5\,\Omega,$$

and the maximum load resistance is

$$R_{Lmax} = \frac{V_O}{I_{Omin}} = \frac{5}{1} = 5\,\Omega.$$

The minimum, nominal, and maximum values of the dc transfer function are

$$M_{V\,DCmin} = \frac{V_O}{V_{Imax}} = \frac{5}{16} = 0.3125,$$

$$M_{V\,DCnom} = \frac{V_O}{V_{Inom}} = \frac{5}{12} = 0.4167,$$

$$M_{V\,DCmax} = \frac{V_O}{V_{Imin}} = \frac{5}{8} = 0.625.$$

Let us assume the converter efficiency $\eta = 80\,\%$. Hence, the minimum, nominal, a maximum values of the duty cycle are

$$D_{min} = \frac{M_{V\,DCmin}}{\eta} = \frac{0.3125}{0.8} = 0.3906,$$

$$D_{nom} = \frac{M_{V\,DCnom}}{\eta} = \frac{0.4167}{0.8} = 0.5209,$$

$$D_{max} = \frac{M_{V\,DCmax}}{\eta} = \frac{0.625}{0.8} = 0.7813.$$

The minimum inductance is

$$L_{min} = \frac{R_{Lmax}(1 - D_{min})}{2f_s} = \frac{5 \times (1 - 0.3906)}{2 \times 100 \times 10^3} = 15.235\,\mu H.$$

Pick $L = 20\,\mu H$.

The ripple voltage is

$$V_r = \frac{V_O}{100} = \frac{5}{100} = 50\,mV.$$

The maximum peak-to-peak ripple of the inductor current is

$$\Delta i_{Lmax} = \frac{V_O(1 - D_{min})}{f_s L} = \frac{5 \times (1 - 0.3906)}{100 \times 10^3 \times 20 \times 10^{-6}} = 1.5235\,A.$$

The maximum value of the ESR of the filter capacitor is

$$r_{Cmax} = \frac{V_r}{\Delta i_{Lmax}} = \frac{0.05}{1.5235} = 32.82\,\text{m}\Omega.$$

Let $r_C = 30\,\text{m}\Omega$. The minimum filter capacitance is

$$C_{min} = \max\left\{\frac{D_{max}}{2f_s r_C}, \frac{1 - D_{min}}{2f_s r_C}\right\} = \max\left\{\frac{0.7813}{2f_s r_C}, \frac{1 - 0.3906}{2f_s r_C}\right\}$$

$$= \frac{0.7813}{2 \times 100 \times 10^3 \times 30 \times 10^{-3}} = 130\,\mu\text{F}.$$

Pick $C = 200\,\mu\text{F}/10\,\text{V}/30\,\text{m}\Omega$.

The transistor and diode voltage stresses are

$$V_{SMmax} = V_{DMmax} = V_{Imax} = 16\,\text{V}.$$

The transistor and diode current stresses are

$$I_{SMmax} = I_{DMmax} = I_{Omax} + \frac{\Delta i_{Lmax}}{2}$$

$$= 10 + \frac{1.5235}{2} = 10.762\,\text{A}.$$

The maximum conduction power loss in the power MOSFET is

$$P_{rDS} = r_{DS} D_{max} I_{Omax}^2 = 10 \times 10^{-3} \times 0.7813 \times 10^2 = 0.7813\,\text{W}.$$

However, the power losses and the efficiency will be calculated here at $V_{Imax} = 16\,\text{V}$, $R_{Lmin} = 0.5\,\Omega$, and $D_{min} = 0.3906$.

$$P_{rDS} = r_{DS} D_{min} I_{Omax}^2 = 10 \times 10^{-3} \times 0.3906 \times 10^2 = 0.3906\,\text{W}.$$

Assuming $C_o = 200\,\text{pF}$, the switching loss is

$$P_{sw} = f_s C_o V_{Imax}^2 = 100 \times 10^3 \times 200 \times 10^{-12} \times 16^2 = 5\,\text{mW}.$$

The total power loss in the MOSFET at $V_{Imax} = 16\,\text{V}$ is

$$P_{FET} = P_{rDS} + \frac{P_{sw}}{2} = 0.3906 + \frac{0.005}{2} = 0.3931\,\text{W}.$$

The power loss in the diode due to the offset voltage V_F is

$$P_{VF} = V_F(1 - D_{min})I_{Omax} = 0.3 \times (1 - 0.3906) \times 10 = 1.828\,\text{W}$$

and the power loss in the diode due to the forward resistance R_F is

$$P_{RF} = R_F(1 - D_{min})I_{Omax}^2$$

$$= 20 \times 10^{-3} \times (1 - 0.3906) \times 10^2 = 1.219\,\text{W}.$$

Hence, the conduction power loss in the diode is

$$P_D = P_{VF} + P_{RF} = 1.828 + 1.219 = 3.047\,\text{W}.$$

The power loss in the inductor is

$$P_{rL} = r_{L(dc)} I_{Omax}^2 = 0.05 \times 10^2 = 5 \,\text{W}.$$

The power loss in the filter capacitor is

$$P_{rC} = \frac{r_C \, \Delta i_{Lmax}^2}{12} = \frac{0.03 \times 1.5235^2}{12} = 6 \,\text{mW}.$$

The total power loss is

$$P_{LS} = P_{rDS} + P_{sw} + P_D + P_{rL} + P_{rC}$$

$$= 0.3906 + 0.005 + 3.047 + 5 + 0.006 = 8.56 \,\text{W}.$$

Hence, the converter efficiency is

$$\eta = \frac{P_{Omax}}{P_{Omax} + P_{LS}} = \frac{50}{50 + 8.56} = 85.37 \,\%.$$

2.12 Design a universal buck PWM converter to meet the following specification
$V_{Imin} = 85\sqrt{2} \,\text{V}$, $V_{Imax} = 264\sqrt{2} \,\text{V}$, $V_O = 48 \,\text{V}$, $I_O = 0.2$ to 2 A, $V_r/V_O \leq 1\,\%$, r_L
$1 \,\Omega$, $r_{DS} = 1 \,\Omega$, $C_o = 100 \,\text{pF}$, $V_F = 0.7 \,\text{V}$, and $R_F = 25 \,\text{m}\Omega$.

The maximum output power is

$$P_{Omax} = V_O I_{Omax} = 48 \times 2 = 96 \,\text{W},$$

and the minimum output power is

$$P_{Omin} = V_O I_{Omin} = 48 \times 0.2 = 9.6 \,\text{W}.$$

The minimum load resistance is

$$R_{Lmin} = \frac{V_O}{I_{Omax}} = \frac{48}{2} = 24 \,\Omega,$$

and the maximum load resistance is

$$R_{Lmax} = \frac{V_O}{I_{Omin}} = \frac{48}{0.2} = 240 \,\Omega.$$

The minimum dc input voltage is

$$V_{Imin} = 85\sqrt{2} = 120.21 \,\text{V},$$

and the maximum dc input voltage is

$$V_{Imax} = 264\sqrt{2} = 373.35 \,\text{V}.$$

The maximum dc voltage transfer function is

$$M_{V \, DCmax} = \frac{V_O}{V_{Imin}} = \frac{48}{120.21} = 0.4,$$

and the minimum dc voltage transfer function is

$$M_{V \, DCmin} = \frac{V_O}{V_{Imax}} = \frac{48}{373.35} = 0.1286.$$

Assume the converter efficiency $\eta = 90\%$. The minimum duty cycle is

$$D_{min} = \frac{M_{V\,DCmin}}{\eta} = \frac{0.1286}{0.9} = 0.143,$$

and the maximum duty cycle is

$$D_{max} = \frac{M_{V\,DCmax}}{\eta} = \frac{0.4}{0.9} = 0.4444.$$

The minimum inductance for CCM is

$$L_{min} = \frac{R_{Lmax}(1 - D_{min})}{2f_s} = \frac{240 \times (1 - 0.143)}{2 \times 200 \times 10^3} = 0.5142\,\text{mH}.$$

Pick $L = 0.6\,\text{mH}$. The maximum inductor ripple current is

$$\Delta i_{Lmax} = \frac{V_O(1 - D_{min})}{f_s L} = \frac{48 \times (1 - 0.143)}{200 \times 10^3 \times 0.6 \times 10^{-3}} = 0.3428\,\text{A}.$$

The ripple output voltage is

$$V_r = 0.01 V_O = 0.01 \times 48 = 0.48\,\text{V}.$$

The maximum filter capacitor ESR is

$$r_{Cmax} = \frac{V_r}{\Delta i_{Lmax}} = \frac{0.48}{0.3428} = 1.4\,\Omega.$$

Pick $r_C = 1\,\Omega$. The minimum filter capacitance is

$$C_{min} = \max\left\{ \frac{D_{max}}{2f_s r_C}, \frac{1 - D_{min}}{2f_s r_C} \right\} = \max\left\{ \frac{0.4444}{2f_s r_C}, \frac{1 - 0.143}{2f_s r_C} \right\}$$

$$= \frac{0.857}{2 \times 200 \times 10^3} = 2.1425\,\mu\text{F}.$$

Pick $C = 5\,\mu\text{F}/100\,\text{V}/1\,\Omega$.

The maximum switch and diode voltage is

$$V_{SMmax} = V_{DMmax} = V_{Imax} = 373.35\,\text{V}$$

and the maximum switch and diode current is

$$I_{SMmax} = I_{DMmax} = I_{Omax} + \frac{\Delta i_{Lmax}}{2} = 2 + \frac{0.3428}{2} = 2.1714\,\text{A}.$$

The power transistor conduction loss at $V_{Imax} = 373.35\,\text{V}$ and $I_{Omax} = 2\,\text{A}$ is

$$P_{rDS} = r_{DS} D_{min} I_{Omax}^2 = 1 \times 0.143 \times 2^2 = 0.572\,\text{W}.$$

The switching power loss is

$$P_{sw} = f_s C_o V_{Imax}^2 = 200 \times 10^3 \times 100 \times 10^{-12} \times 373.35^2 = 2.79\,\text{W}.$$

The total power loss in the MOSFET is

$$P_{MOS} = P_{rDS} + \frac{P_{sw}}{2} = 0.572 + \frac{2.79}{2} = 1.967\,\text{W}.$$

The power loss in the diode due to V_F is

$$P_{VF} = V_F(1 - D_{min})I_{Omax} = 0.7 \times (1 - 0.143) \times 2 = 1.2 \, \text{W},$$

and the power loss in the diode due to R_F is

$$P_{RF} = R_F(1 - D_{min})I_{Omax}^2 = 0.025 \times (1 - 0.143) \times 2^2 = 0.0857 \, \text{W}.$$

Hence, the total conduction power loss in the diode is

$$P_D = P_{VF} + P_{RF} = 1.2 + 0.0857 = 1.2857 \, \text{W}.$$

The power loss in the inductor is

$$P_{rL} = r_L I_{Omax}^2 = 1 \times 2^2 = 4 \, \text{W}.$$

The power loss in the filter capacitor is

$$P_{rC} = \frac{r_C \Delta i_{Lmax}^2}{12} = \frac{1 \times 0.3428^2}{12} = 0.0098 \, \text{W}.$$

Thus, the total power loss is

$$P_{LS} = P_{rDS} + P_{sw} + P_D + P_{rL} + P_{rC}$$

$$= 0.572 + 2.79 + 1.2857 + 4 + 0.0098 = 8.657 \, \text{W},$$

resulting in the efficiency

$$\eta = \frac{P_{Omax}}{P_{Omax} + P_{LS}} = \frac{96}{96 + 8.657} = 91.728 \, \%.$$

2.13 A buck converter has the following specifications: $V_I = 4$ to 6 V, $V_O = 3$ V, $I_O = 0$ ⬤
5 A, and $V_r/V_O \leq 2\,\%$. Assume $\eta = 0.9$. Find L, C, and r_C.

The minimum load resistance is

$$R_{Lmin} = \frac{V_O}{I_{Omax}} = \frac{3}{5} = 0.6 \, \Omega.$$

The maximum dc voltage transfer function is

$$M_{V\,DCmax} = \frac{V_O}{V_{Imin}} = \frac{3}{4} = 0.75.$$

Assume the converter efficiency $\eta = 90\,\%$. Then the maximum duty cycle at the bound
ary between CCM and DCM at full load is

$$D_{Bmax} = \frac{M_{V\,DCmax}}{\eta} = \frac{0.75}{0.9} = 0.8333.$$

Hence, the maximum inductance required for DCM operation is

$$L_{max} = \frac{R_{Lmin}(1 - D_{Bmax})}{2f_s} = \frac{0.6 \times (1 - 0.8333)}{2 \times 250 \times 10^3} = 0.2 \, \mu\text{H}.$$

Pick $L = 0.15 \, \mu\text{H}$.

The minimum dc voltage transfer function is

$$M_{V\,DCmin} = \frac{V_O}{V_{Imax}} = \frac{3}{6} = 0.5.$$

The minimum duty cycle at full load is

$$D_{min} = \sqrt{\frac{2f_s LM_{V\,DCmin}^2}{\eta R_{Lmin}(1 - M_{V\,DCmin})}}$$

$$= \sqrt{\frac{2 \times 250 \times 10^3 \times 0.15 \times 10^{-6} \times 0.5^2}{0.9 \times 0.6 \times (1 - 0.5)}} = 0.2635.$$

Thus, the maximum peak inductor current is

$$\Delta i_{Lmax} = \frac{D_{min}(V_{Imax} - V_O)}{f_s L} = \frac{0.2635 \times (6 - 3)}{250 \times 10^3 \times 0.15 \times 10^{-6}} = 21.08\,\text{A}.$$

The ripple voltage is

$$V_r = 0.02 V_O = 0.02 \times 3 = 0.06\,\text{V}.$$

Hence, the maximum filter capacitor ESR is

$$r_{Cmax} = \frac{V_r}{\Delta i_{Lmax}} = \frac{0.06}{21.08} = 2.85\,\text{m}\Omega.$$

Pick $r_C = 2.5\,\text{m}\Omega$.

The maximum duty cycle at full load is

$$D_{max} = \sqrt{\frac{2f_s LM_{V\,DCmax}^2}{\eta R_{Lmin}(1 - M_{V\,DCmax})}}$$

$$= \sqrt{\frac{2 \times 250 \times 10^3 \times 0.15 \times 10^{-6} \times 0.75^2}{0.9 \times 0.6 \times (1 - 0.75)}} = 0.559.$$

The minimum filter capacitance is

$$C_{min} = \max\left\{ \frac{D_{max}}{2f_s r_C}, \frac{1 - D_{min}}{2f_s r_C} \right\} = \max\left\{ \frac{0.559}{2f_s r_C}, \frac{1 - 0.2635}{2f_s r_C} \right\}$$

$$= \frac{1 - 0.2635}{2 \times 250 \times 10^3 \times 2.5 \times 10^{-3}} = 589.2\,\mu\text{F}.$$

Pick $C = 1\,\text{mF}/10\,\text{V}/2.5\,\text{m}\Omega$.

.14 A buck PWM converter has $V_I = 270\,\text{V} \pm 5\%$, $V_O = 28\,\text{V}$, $I_O = 0$ to $15\,\text{A}$, $V_r/V_O \leq 5\%$, $r_{L(dc)} = 0.05\,\Omega$, $r_C = 0.037\,\Omega$, $r_{DS} = 0.3\,\Omega$, $C_o = 150\,\text{pF}$, $V_F = 0.8\,\text{V}$, $R_F = 17.1\,\text{m}\Omega$, and $f_s = 100\,\text{kHz}$. Find L, C, I_{SM}, V_{SM}, P_{LS}, and η. Assume the initial efficiency $\eta = 90\%$ at full power.

The minimum load resistance is

$$R_{Lmin} = \frac{V_O}{I_{Omax}} = \frac{28}{15} = 1.867\,\Omega.$$

The minimum input voltage is

$$V_{Imin} = 270 - 0.05 \times 270 = 256.5 \, \text{V},$$

and the maximum input voltage is

$$V_{Imax} = 270 + 0.05 \times 270 = 283.5 \, \text{V}.$$

The minimum dc voltage transfer function is

$$M_{V\,DCmin} = \frac{V_O}{V_{Imax}} = \frac{28}{283.5} = 0.09877,$$

and the maximum dc voltage transfer function is

$$M_{V\,DCmax} = \frac{V_O}{V_{Imin}} = \frac{28}{256.5} = 0.1092.$$

Assuming $\eta = 90\,\%$, the maximum duty cycle at the CCM/DCM boundary at full load

$$D_{Bmax} = \frac{M_{V\,DCmax}}{\eta} = \frac{0.1092}{0.9} = 0.1213.$$

Hence, the maximum inductance required for DCM operation is

$$L_{max} = \frac{R_{Lmin}(1 - D_{Bmax})}{2f_s} = \frac{1.867 \times (1 - 0.1213)}{2 \times 100 \times 10^3} = 8.2 \, \mu\text{H}.$$

Pick $L = 6 \, \mu\text{H}$.

Assuming $\eta = 0.9$, the minimum duty cycle at full load is

$$D_{min} = \sqrt{\frac{2f_s L M_{V\,DCmin}^2}{\eta R_{Lmin}(1 - M_{V\,DCmin})}}$$

$$= \sqrt{\frac{2 \times 100 \times 10^3 \times 6 \times 10^{-6} \times 0.09877^2}{0.9 \times 1.867 \times (1 - 0.09877)}} = 0.0879,$$

and the maximum duty cycle

$$D_{max} = \sqrt{\frac{2f_s L M_{V\,DCmax}^2}{\eta R_{Lmin}(1 - M_{V\,DCmax})}}$$

$$= \sqrt{\frac{2 \times 100 \times 10^3 \times 6 \times 10^{-6} \times 0.1092^2}{0.9 \times 1.867 \times (1 - 0.1092)}} = 0.0977.$$

Thus, the maximum peak inductor current is

$$\Delta i_{Lmax} = \frac{D_{min}(V_{Imax} - V_O)}{f_s L} = \frac{0.0879 \times (283.5 - 28)}{100 \times 10^3 \times 6 \times 10^{-6}} = 37.431 \, \text{A}.$$

The ripple voltage is

$$V_r = 0.05 V_O = 0.05 \times 28 = 1.4 \, \text{V}.$$

Hence, the maximum filter capacitor ESR is

$$r_{Cmax} = \frac{V_r}{\Delta i_{Lmax}} = \frac{1.4}{37.431} = 37.4\,m\Omega.$$

Pick $r_C = 20\,m\Omega$.

The minimum filter capacitance is

$$C_{min} = \max\left\{\frac{D_{max}}{2f_s r_C}, \frac{1 - D_{min}}{2f_s r_C}\right\} = \max\left\{\frac{0.0977}{2f_s r_C}, \frac{1 - 0.0879}{2f_s r_C}\right\}$$

$$= \frac{0.9121}{2 \times 100 \times 10^3 \times 20 \times 10^{-3}} = 0.228\,mF.$$

Pick $C = 0.25\,mF/50\,V/20\,m\Omega$.

Chapter 3
Boost PWM DC-DC Converter

1 Derive an expression for the dc voltage transfer function M_{VDC} of a lossless boost converter using the steady-state condition for the inductor current.

In steady state,

$$i_L(T) = i_L(0).$$

Therefore, the increase in the inductor current during the switch on-time interval is equal to the decrease in the inductor current during the switch off-time interval. During the switch on-time ($0 < t \leq DT$),

$$v_L = V_I,$$

and the inductor current is

$$i_L = \frac{1}{L} \int_0^t V_I \, dt + i_L(0) = \frac{V_I}{L} t + i_L(0).$$

At $t = DT$,

$$i_L(DT) = \frac{V_I DT}{L} + i_L(0),$$

and the inductor current increase during the switch on-time is

$$\Delta i_{L(on)} = i_L(DT) - i_L(0) = \frac{V_I DT}{L} = \frac{V_I D}{f_s L}.$$

During the switch off-time ($DT < t \leq T$),

$$v_L = V_I - V_O < 0,$$

resulting in the inductor current

$$i_L = \frac{1}{L} \int_{DT}^t (V_I - V_O) \, dt = \frac{V_I - V_O}{L} (t - DT) + i_L(DT).$$

Pulse-width Modulated DC–DC Converters – Solutions Manual   Marian K. Kazimierczuk
2008 John Wiley & Sons, Ltd

At $t = T$,

$$i_L(T) = \frac{V_I - V_O}{L}(T - DT) + i_L(DT) = \frac{V_I - V_O}{L}(1 - D)T + i_L(DT).$$

Thus, the inductor current decrease during the switch off-time is

$$\Delta i_{L(off)} = i_L(DT) - i_L(T) = -\frac{V_I - V_O}{L}(1 - D)T$$

$$= -\frac{(V_I - V_O)(1 - D)}{f_s L}.$$

Equating the inductor current increase and decrease, one obtains

$$\Delta i_{L(on)} = \Delta i_{L(off)}$$

which gives

$$\frac{V_I DT}{L} = -\frac{V_I - V_O}{L}(1 - D)T.$$

Rearrangement of this equation gives

$$M_{V\,DC} = \frac{V_O}{V_I} = \frac{1}{1 - D}.$$

3.2 A boost PWM converter has the following data: $V_I = 125$ to $350\,\text{V}$, $V_O = 380\,\text{V}$, P_O 6.8 to 68 W, and $f_s = 50\,\text{kHz}$. Compute the voltage and current stresses of the transistor and the diode.

The transistor and diode voltage stress is

$$V_{SM} = V_{DM} = V_O = 380\,\text{V}.$$

The maximum load current is

$$I_{Omax} = \frac{P_{Omax}}{V_O} = \frac{68}{380} = 0.179\,\text{A}.$$

Because the dc voltage transfer function is given by

$$M_{V\,DC} = \frac{V_O}{V_I} = \frac{1}{1 - D},$$

the minimum duty cycle is

$$D_{min} = 1 - \frac{V_{Imax}}{V_O} = 1 - \frac{350}{380} = 0.0789$$

and the maximum duty cycle is

$$D_{max} = 1 - \frac{V_{Imin}}{V_O} = 1 - \frac{125}{380} = 0.671.$$

The maximum load resistance is

$$R_{Lmax} = \frac{V_O^2}{P_{Omin}} = \frac{380^2}{6.8} = 21.235\,\text{k}\Omega.$$

Since the duty cycle D varies between 0.0789 and 0.671, the minimum inductance for CCM operation is

$$L_{min} = \frac{2}{27} \frac{R_{Lmax}}{f_s} = \frac{2}{27} \frac{21.235 \times 10^3}{50 \times 10^3} = 31.459 \, \text{mH}.$$

Pick $L = 32 \, \text{mH}$.

The maximum peak-to-peak ripple of the inductor current occurs at $D = 0.5$ and is given by

$$\Delta i_{Lmax} = \frac{V_O}{4 f_s L} = \frac{380}{4 \times 50 \times 10^3 \times 32 \times 10^{-3}} = 0.059 \, \text{A}.$$

The transistor and diode current stress is

$$I_{SMmax} = I_{DMmax} = \frac{I_{Omax}}{1 - D_{max}} + \frac{\Delta i_{Lmax}}{2} = \frac{0.179}{1 - 0.671} + \frac{0.059}{2}$$

$$= 0.544 + 0.03 = 0.574 \, \text{A}.$$

.3 A boost PWM converter has the following data: $V_I = 8$ to $16 \, \text{V}$, $V_O = 24 \, \text{V}$, $I_O = 0.2$ to $2 \, \text{A}$, and $f_s = 200 \, \text{kHz}$. Calculate the minimum inductance required for the converter to operate in CCM. Assume $\eta = 90 \, \%$.

The maximum load resistance is

$$R_{Lmax} = \frac{V_O}{I_{Omin}} = \frac{24}{0.2} = 120 \, \Omega.$$

The dc voltage transfer function is

$$M_{V \, DC} = \frac{V_O}{V_I} = \frac{\eta}{1 - D}.$$

The minimum duty cycle is

$$D_{min} = 1 - \eta \frac{V_{Imax}}{V_O} = 1 - 0.9 \times \frac{16}{24} = 0.4,$$

and the maximum duty cycle is

$$D_{max} = 1 - \eta \frac{V_{Imin}}{V_O} = 1 - 0.9 \times \frac{8}{24} = 0.7.$$

The maximum load boundary current occurs at $D_{min} = D_m = 1/3$. The minimum inductance is then

$$L_{min} = \frac{2}{27} \frac{R_{Lmax}}{f_s} = \frac{2}{27} \frac{120}{0.2 \times 10^6} = 44.44 \, \mu\text{H}.$$

Pick $L = 50 \, \mu\text{H}$.

.4 A boost PWM converter has the following data: $V_I = 8$ to $12 \, \text{V}$, $V_O = 24 \, \text{V}$, $I_O = 0.2$ to $2 \, \text{A}$, and $f_s = 200 \, \text{kHz}$. Calculate the minimum inductance required for the converter to operate in CCM. Assume $\eta = 90 \, \%$.

The minimum load resistance is

$$R_{Lmax} = \frac{V_O}{I_{Omin}} = \frac{24}{0.2} = 120\,\Omega.$$

The minimum duty cycle is

$$D_{min} = 1 - \eta\frac{V_{Imax}}{V_O} = 1 - 0.9 \times \frac{12}{24} = 0.45,$$

and the maximum duty cycle is

$$D_{max} = 1 - \eta\frac{V_{Imin}}{V_O} = 1 - 0.9 \times \frac{8}{24} = 0.7.$$

Notice that $D_{min} > D_m = 1/3$. Therefore,

$$L_{min} = \frac{R_{Lmax}D_{min}(1 - D_{min})^2}{2f_s}$$

$$= \frac{120 \times 0.45 \times (1 - 0.45)^2}{2 \times 0.2 \times 10^6} = 40.8\,\mu\text{H}.$$

Pick $L = 47\,\mu\text{H}$.

3.5 A boost PWM converter is operated in CCM at $V_I = 14\,\text{V}$ and $V_O = 28\,\text{V}$. Find the required duty cycle D for the converter efficiency (a) $\eta = 100\,\%$ and (b) $\eta = 80\,\%$.

The dc voltage transfer function of the boost converter is

$$M_{V\,DC} = \frac{V_O}{V_I} = \frac{\eta}{1 - D}.$$

Hence, the required duty cycle is

$$D = 1 - \frac{\eta}{M_{V\,DC}} = 1 - \frac{\eta V_I}{V_O}.$$

For $\eta = 100\,\%$,

$$D = 1 - \frac{\eta V_I}{V_O} = 1 - \frac{1 \times 14}{28} = 0.5.$$

For $\eta = 80\,\%$,

$$D = 1 - \frac{\eta V_I}{V_O} = 1 - \frac{0.8 \times 14}{28} = 0.6.$$

3.6 A boost PWM converter employs a power MOSFET with an on-resistance $r_{DS} = 0.02\,\Omega$. The load current is $I_O = 10\,\text{A}$. Calculate the transistor conduction loss at $D = 0.1, 0.2, 0.3, 0.4, 0.5, 0.6, 0.7, 0.8,$ and 0.9.

At $D = 0.1$,

$$P_{rDS} = \frac{Dr_{DS}I_O^2}{(1 - D)^2} = \frac{0.1 \times 0.02 \times 10^2}{(1 - 0.1)^2} = 0.247\,\text{W}.$$

At $D = 0.2$,
$$P_{rDS} = \frac{Dr_{DS}I_O^2}{(1-D)^2} = \frac{0.2 \times 0.02 \times 10^2}{(1-0.2)^2} = 0.625 \text{ W}.$$

At $D = 0.3$,
$$P_{rDS} = \frac{Dr_{DS}I_O^2}{(1-D)^2} = \frac{0.3 \times 0.02 \times 10^2}{(1-0.3)^2} = 1.224 \text{ W}.$$

At $D = 0.4$,
$$P_{rDS} = \frac{Dr_{DS}I_O^2}{(1-D)^2} = \frac{0.4 \times 0.02 \times 10^2}{(1-0.4)^2} = 2.222 \text{ W}.$$

At $D = 0.5$,
$$P_{rDS} = \frac{Dr_{DS}I_O^2}{(1-D)^2} = \frac{0.5 \times 0.02 \times 10^2}{(1-0.5)^2} = 4 \text{ W}.$$

At $D = 0.6$,
$$P_{rDS} = \frac{Dr_{DS}I_O^2}{(1-D)^2} = \frac{0.6 \times 0.02 \times 10^2}{(1-0.6)^2} = 7.5 \text{ W}.$$

At $D = 0.7$,
$$P_{rDS} = \frac{Dr_{DS}I_O^2}{(1-D)^2} = \frac{0.7 \times 0.02 \times 10^2}{(1-0.7)^2} = 15.556 \text{ W}.$$

At $D = 0.8$,
$$P_{rDS} = \frac{Dr_{DS}I_O^2}{(1-D)^2} = \frac{0.8 \times 0.02 \times 10^2}{(1-0.8)^2} = 40 \text{ W}.$$

At $D = 0.9$,
$$P_{rDS} = \frac{Dr_{DS}I_O^2}{(1-D)^2} = \frac{0.9 \times 0.02 \times 10^2}{(1-0.9)^2} = 180 \text{ W}.$$

It can be seen that the transistor conduction loss P_{rDS} increases very rapidly with increasing duty cycle D, especially at higher values of D.

.7 A boost PWM converter employs a diode with a forward resistance $R_F = 0.02 \, \Omega$. The load current is $I_O = 10$ A. Calculate the diode conduction loss due to the forward resistance R_F at $D = 0.1$, 0.2, 0.5, 0.8, and 0.9.

At $D = 0.1$,
$$P_{RF} = \frac{R_F I_O^2}{1-D} = \frac{0.02 \times 10^2}{1-0.1} = 2.222 \text{ W}.$$

At $D = 0.2$,
$$P_{RF} = \frac{R_F I_O^2}{1-D} = \frac{0.02 \times 10^2}{1-0.2} = 2.5 \text{ W}.$$

At $D = 0.5$,

$$P_{RF} = \frac{R_F I_O^2}{1 - D} = \frac{0.02 \times 10^2}{1 - 0.5} = 4\,\text{W}.$$

At $D = 0.8$,

$$P_{RF} = \frac{R_F I_O^2}{1 - D} = \frac{0.02 \times 10^2}{1 - 0.8} = 10\,\text{W}.$$

At $D = 0.9$,

$$P_{RF} = \frac{R_F I_O^2}{1 - D} = \frac{0.02 \times 10^2}{1 - 0.9} = 20\,\text{W}.$$

It can be seen that the diode conduction loss P_{RF} associated with the diode forward resistance R_F increases rapidly with increasing duty cycle D, especially at higher value of D.

3.8 A boost PWM converter employs an inductor with a dc resistance $r_L = 0.02\,\Omega$. The load current is $I_O = 10\,\text{A}$. Calculate the inductor loss at $D = 0.1, 0.2, 0.3, 0.4, 0.5, 0.6, 0.7$ 0.8, and 0.9.

At $D = 0.1$,

$$P_{rL} = \frac{r_L I_O^2}{(1 - D)^2} = \frac{0.02 \times 10^2}{(1 - 0.1)^2} = 2.469\,\text{W}.$$

At $D = 0.2$,

$$P_{rL} = \frac{r_L I_O^2}{(1 - D)^2} = \frac{0.02 \times 10^2}{(1 - 0.2)^2} = 3.125\,\text{W}.$$

At $D = 0.3$,

$$P_{rL} = \frac{r_L I_O^2}{(1 - D)^2} = \frac{0.02 \times 10^2}{(1 - 0.3)^2} = 4.082\,\text{W}.$$

At $D = 0.4$,

$$P_{rL} = \frac{r_L I_O^2}{(1 - D)^2} = \frac{0.02 \times 10^2}{(1 - 0.4)^2} = 5.556\,\text{W}.$$

At $D = 0.5$,

$$P_{rL} = \frac{r_L I_O^2}{(1 - D)^2} = \frac{0.02 \times 10^2}{(1 - 0.5)^2} = 8\,\text{W}.$$

At $D = 0.6$,

$$P_{rL} = \frac{r_L I_O^2}{(1 - D)^2} = \frac{0.02 \times 10^2}{(1 - 0.6)^2} = 12.5\,\text{W}.$$

At $D = 0.7$,

$$P_{rL} = \frac{r_L I_O^2}{(1 - D)^2} = \frac{0.02 \times 10^2}{(1 - 0.7)^2} = 22.222\,\text{W}.$$

At $D = 0.8$,

$$P_{rL} = \frac{r_L I_O^2}{(1-D)^2} = \frac{0.02 \times 10^2}{(1-0.8)^2} = 50\,\text{W}.$$

At $D = 0.9$,

$$P_{iL} - \frac{r_L I_O^2}{(1-D)^2} = \frac{0.02 \times 10^2}{(1-0.9)^7} = 200\,\text{W}.$$

It can be seen that the inductor power loss P_{rL} increases very rapidly with increasing duty cycle D, especially at higher values of D.

9   For the boost converter with $V_I = 8$ to $16\,\text{V}$, $V_O = 24\,\text{V}$, $I_O = 0.2$ to $2\,\text{A}$, and $f_s = 200\,\text{kHz}$, find the maximum inductance required to maintain the converter in DCM.

The minimum load resistance is

$$R_{Lmin} = \frac{V_O}{I_{Omax}} = \frac{24}{2} = 12\,\Omega.$$

Since the dc voltage transfer function at the CCM/DCM boundary is the same as in CCM, we obtain

$$M_{V\,DC} = \frac{V_O}{V_I} = \frac{1}{1-D},$$

the minimum duty cycle is

$$D_{min} = 1 - \frac{1}{M_{V\,DCmin}} = 1 - \frac{V_{Imax}}{V_O} = 1 - \frac{16}{24} = \frac{1}{3},$$

and the maximum duty cycle is

$$D_{max} = 1 - \frac{1}{M_{V\,DCmax}} = 1 - \frac{V_{Imin}}{V_O} = 1 - \frac{8}{24} = \frac{2}{3}.$$

Because $D \leq 1/3$,

$$L_{max} = \frac{R_{Lmin}D_{max}(1-D_{max})^2}{2f_s} = \frac{12 \times \frac{2}{3} \times \left(1 - \frac{2}{3}\right)^2}{2 \times 0.2 \times 10^6} = 2.22\,\mu\text{H}.$$

Chapter 4

Buck-boost PWM DC-DC Converter

1 A buck-boost PWM converter has $V_I = 127$ to $187\,V$, $V_O = 48\,V$, $I_O = 1$ to $2\,A$, $\eta = 100\,\%$, and $f_s = 50\,kHz$. Find the minimum inductance required for CCM operation.

The minimum load resistance is

$$R_{Lmin} = \frac{V_O}{I_{Omax}} = \frac{48}{2} = 24\,\Omega,$$

and the maximum load resistance is

$$R_{Lmax} = \frac{V_O}{I_{Omin}} = \frac{48}{1} = 48\,\Omega.$$

The minimum dc voltage transfer function is

$$M_{V\,DCmin} = \frac{D_{min}}{1 - D_{min}} = \frac{V_O}{V_{Imax}} = \frac{48}{187} = 0.2567,$$

and the maximum dc voltage transfer function is

$$M_{V\,DCmax} = \frac{D_{max}}{1 - D_{max}} = \frac{V_O}{V_{Imin}} = \frac{48}{127} = 0.378.$$

Hence,

$$D_{min} = \frac{M_{V\,DCmin}}{M_{V\,DCmin} + \eta} = \frac{0.2567}{0.2567 + 1} = 0.2043$$

and

$$D_{max} = \frac{M_{V\,DCmax}}{M_{V\,DCmax} + \eta} = \frac{0.378}{0.378 + 1} = 0.2743.$$

The minimum inductance is

$$L_{min} = \frac{R_{Lmax}(1 - D_{min})^2}{2f_s} = \frac{48 \times (1 - 0.2043)^2}{2 \times 50 \times 10^3} = 304\,\mu H.$$

Pick $L = 330\,\mu H$.

4.? A buck-boost PWM converter has $V_I = 127$ to 187 V, $V_O = 48$ V, $I_O = 1$ to 2 A, η = 85 %, and $f_s = 50$ kHz. Find the minimum inductance required for CCM operation.

The maximum load resistance is

$$R_{Lmax} = \frac{V_O}{I_{Omin}} = \frac{48}{1} = 48\,\Omega.$$

The minimum dc voltage transfer function is

$$M_{V\,DCmin} = \frac{D_{min}}{1 - D_{min}} = \frac{V_O}{V_{Imax}} = \frac{48}{187} = 0.2567,$$

and the maximum dc voltage transfer function is

$$M_{V\,DCmax} = \frac{D_{max}}{1 - D_{max}} = \frac{V_O}{V_{Imin}} = \frac{48}{127} = 0.378,$$

from which

$$D_{min} = \frac{M_{V\,DCmin}}{M_{V\,DCmin} + \eta} = \frac{0.2567}{0.2567 + 0.85} = 0.23195$$

and

$$D_{max} = \frac{M_{V\,DCmax}}{M_{V\,DCmax} + \eta} = \frac{0.378}{0.378 + 0.85} = 0.3078.$$

The minimum inductance is

$$L_{min} = \frac{R_{Lmax}(1 - D_{min})^2}{2f_s} = \frac{48 \times (1 - 0.23195)^2}{2 \times 50 \times 10^3} = 283.15\,\mu\text{H}.$$

Pick $L = 330\,\mu\text{H}$. Thus, the lossy converter requires a lower L_{min} for CCM. Henc calculations of L_{min} with $\eta = 100\,\%$ will always ensure CCM operation.

4.3 A buck-boost PWM converter (given in Problem 4.1) has $V_I = 127$ to 187 V, $V_O = 48$ $I_O = 1$ to 2 A, $L = 420\,\mu\text{H}$, and $f_s = 50$ kHz. Find the voltage and current stresses of t transistor and the diode.

The voltage stress of the transistor and the diode is

$$V_{SMmax} = V_{DMmax} = V_{Imax} + V_O = 187 + 48 = 235\,\text{V}.$$

The maximum peak-to-peak ripple current of the inductor is

$$\Delta i_{Lmax} = \frac{V_O(1 - D_{min})}{f_s L} = \frac{48 \times (1 - 0.2043)}{50 \times 10^3 \times 420 \times 10^{-6}} = 1.819\,\text{A}.$$

The maximum dc input current is

$$I_{Imax} = M_{V\,DCmax}I_{Omax} = 0.378 \times 2 = 0.756\,\text{A}.$$

Hence, the current stress of the transistor and the diode is

$$I_{SMmax} = I_{DMmax} = I_{Imax} + I_{Omax} + \frac{\Delta i_{Lmax}}{2}$$

$$= 0.756 + 2 + \frac{1.819}{2} = 3.666\,\text{A}.$$

4 A buck-boost PWM converter (given in Problem 4.2) has $V_I = 127$ to 187 V, $V_O = 48$ V, $I_O = 1$ to 2 A, $L = 420\,\mu$H, and $f_s = 50$ kHz. Find the filter capacitance and the ESR so that $V_r/V_O \le 1\,\%$.

The minimum dc voltage transfer function is

$$M_{V\,DCmin} = \frac{D_{min}}{1 - D_{min}} = \frac{V_O}{V_{Imax}} = \frac{48}{187} = 0.2567.$$

Assuming $\eta = 0.85$, the minimum duty cycle is

$$D_{min} = \frac{M_{V\,DCmin}}{M_{V\,DCmin} + \eta} = \frac{0.2567}{0.2567 + 0.85} = 0.2043.$$

The maximum peak-to-peak value of the inductor ripple current is

$$\Delta i_{Lmax} = \frac{V_O(1 - D_{min})}{f_s L} = \frac{48 \times (1 - 0.23195)}{50 \times 10^3 \times 420 \times 10^{-6}} = 1.818\,\text{A}.$$

The ripple voltage is

$$V_r = \frac{48}{100} = 0.48\,\text{V}.$$

Let us assume that the ripple voltage across the ESR is $V_{rc} = 0.3$ V. Hence, the ripple voltage across the filter capacitance is

$$V_{Cpp} = V_r - V_{rc} = 0.48 - 0.3 = 0.18\,\text{V}.$$

The minimum load resistance is

$$R_{Lmin} = \frac{V_O}{I_{Omax}} = \frac{48}{2} = 24\,\Omega.$$

The maximum dc voltage transfer function is

$$M_{V\,DCmax} = \frac{D_{max}}{1 - D_{max}} = \frac{V_O}{V_{Imin}} = \frac{48}{127} = 0.378,$$

and the maximum duty cycle is

$$D_{max} = \frac{M_{V\,DCmax}}{M_{V\,DCmax} + \eta} = \frac{0.378}{0.378 + 0.85} = 0.2743.$$

The maximum peak value of the diode current is

$$I_{DMmax} = \frac{I_{Omax}}{1 - D_{max}} + \frac{\Delta i_{Lmax}}{2} = \frac{2}{1 - 0.2743} + \frac{1.756}{2} = 3.665\,\text{A}.$$

Thus, the maximum ESR is

$$r_{Cmax} = \frac{V_{rc}}{I_{DMmax}} = \frac{0.3}{3.665} = 81.85\,\text{m}\Omega.$$

The minimum load resistance is

$$R_{Lmin} = \frac{V_O}{I_{Omax}} = \frac{48}{2} = 24\,\Omega.$$

The minimum filter capacitance is

$$C_{min} = \frac{D_{max}}{f_s R_{Lmin}} \frac{V_O}{V_{Cpp}} = \frac{0.2743}{50 \times 10^3 \times 24} \frac{48}{0.18} = 60.956\,\mu\text{F}.$$

Let $C = 82\,\mu$F/63 V/70 mΩ.

4.5 A buck-boost PWM converter employs a filter capacitor with an ESR $r_C = 10\,\text{m}\Omega$. Th load current is $I_O = 10\,\text{A}$ and the converter operates in CCM. Calculate the power lo in the filter capacitor at $D = 0.1, 0.2, 0.3, 0.4, 0.5, 0.6, 0.7, 0.8$, and 0.9.

At $D = 0.1$,

$$P_{rC} = \frac{Dr_C I_O^2}{1 - D} = \frac{0.1 \times 0.01 \times 10^2}{1 - 0.1} = 0.111\,\text{W}.$$

At $D = 0.2$,

$$P_{rC} = \frac{Dr_C I_O^2}{1 - D} = \frac{0.2 \times 0.01 \times 10^2}{1 - 0.2} = 0.25\,\text{W}.$$

At $D = 0.3$,

$$P_{rC} = \frac{Dr_C I_O^2}{1 - D} = \frac{0.3 \times 0.01 \times 10^2}{1 - 0.3} = 0.429\,\text{W}.$$

At $D = 0.4$,

$$P_{rC} = \frac{Dr_C I_O^2}{1 - D} = \frac{0.4 \times 0.01 \times 10^2}{1 - 0.4} = 0.667\,\text{W}.$$

At $D = 0.5$,

$$P_{rC} = \frac{Dr_C I_O^2}{1 - D} = \frac{0.5 \times 0.01 \times 10^2}{1 - 0.5} = 1\,\text{W}.$$

At $D = 0.6$,

$$P_{rC} = \frac{Dr_C I_O^2}{1 - D} = \frac{0.6 \times 0.01 \times 10^2}{1 - 0.6} = 1.5\,\text{W}.$$

At $D = 0.7$,

$$P_{rC} = \frac{Dr_C I_O^2}{1 - D} = \frac{0.7 \times 0.01 \times 10^2}{1 - 0.7} = 2.333\,\text{W}.$$

At $D = 0.8$,

$$P_{rC} = \frac{Dr_C I_O^2}{1 - D} = \frac{0.8 \times 0.01 \times 10^2}{1 - 0.8} = 4\,\text{W}.$$

At $D = 0.9$,

$$P_{rC} = \frac{Dr_C I_O^2}{1 - D} = \frac{0.9 \times 0.01 \times 10^2}{1 - 0.9} = 9\,\text{W}.$$

4.6 A buck-boost PWM converter has $V_I = 127$ to $187\,\text{V}$, $V_O = 48\,\text{V}$, $I_O = 0$ to $2\,\text{A}$, η $100\,\%$, and $f_s = 50\,\text{kHz}$. Find the maximum inductance required for DCM operation.

The minimum load resistance is

$$R_{Lmin} = \frac{V_O}{I_{Omax}} = \frac{48}{2} = 24\,\Omega.$$

The minimum dc voltage transfer function is

$$M_{V\,DCmin} = \frac{D_{min}}{1 - D_{min}} = \frac{V_O}{V_{Imax}} = \frac{48}{187} = 0.2567,$$

and the maximum dc voltage transfer function is

$$M_{V\,DCmax} = \frac{D_{max}}{1 - D_{max}} = \frac{V_O}{V_{Imin}} = \frac{48}{127} = 0.378,$$

from which

$$D_{min} = \frac{M_{V\,DCmin}}{M_{V\,DCmin} + \eta} = \frac{0.2567}{0.2567 + 1} = 0.2043$$

and

$$D_{max} = \frac{M_{V\,DCmax}}{M_{V\,DCmax} + \eta} = \frac{0.378}{0.378 + 1} = 0.2743.$$

The maximum inductance is

$$L_{max} = \frac{R_{Lmin}(1 - D_{max})^2}{2f_s} = \frac{24 \times (1 - 0.2743)^2}{2 \times 50 \times 10^3} = 126.39\,\mu H.$$

Pick $L = 100\,\mu H$.

7 A buck-boost PWM converter has $V_I = 127$ to $187\,V$, $V_O = 48\,V$, $I_O = 0$ to $2\,A$, $\eta = 85\%$, and $f_s = 50\,kHz$. Find the maximum inductance required for DCM operation.

The minimum load resistance is

$$R_{Lmin} = \frac{V_O}{I_{Omax}} = \frac{48}{2} = 24\,\Omega.$$

The minimum dc voltage transfer function is

$$M_{V\,DCmin} = \frac{D_{min}}{1 - D_{min}} = \frac{V_O}{V_{Imax}} = \frac{48}{187} = 0.2567,$$

and the maximum dc voltage transfer function is

$$M_{V\,DCmax} = \frac{D_{max}}{1 - D_{max}} = \frac{V_O}{V_{Imin}} = \frac{48}{127} = 0.378,$$

from which

$$D_{min} = \frac{M_{V\,DCmin}}{M_{V\,DCmin} + \eta} = \frac{0.2567}{0.2567 + 0.85} = 0.23195$$

and

$$D_{max} = \frac{M_{V\,DCmax}}{M_{V\,DCmax} + \eta} = \frac{0.378}{0.378 + 0.85} = 0.3078.$$

The maximum inductance is

$$L_{max} = \frac{R_{Lmin}(1 - D_{max})^2}{2f_s} = \frac{24 \times (1 - 0.3078)^2}{2 \times 50 \times 10^3} = 115\,\mu H.$$

Pick $L = 100\,\mu H$. Thus, lossy converter requires a lower L_{max}. Therefore, calculations of L_{max} should be done for the lossy converter.

4.8 Derive an expression for the dc voltage transfer function for the lossless buck-boo
PMW converter operating in DCM using the principle of energy conservation.

When the switch is on and the diode is off, the inductor current waveform, which is equ
to the input current waveform, is given by

$$i_I = i_L = \frac{V_I}{L}t, \quad \text{for } 0 < t \leq DT.$$

Hence, the dc component of the converter input current is

$$I_I = \frac{1}{T}\int_0^{DT} i_I \, dt = \frac{V_I}{TL}\int_0^{DT} t \, dt = \frac{V_I D^2 T}{2L} = \frac{V_I D^2}{2 f_s L},$$

resulting in the input power of the converter

$$P_I = I_I V_I = \frac{V_I^2 D^2}{2 f_s L}.$$

The output power of the converter is given by

$$P_O = \frac{V_O^2}{R_L}.$$

For the lossless converter,

$$P_O = P_I,$$

resulting in

$$\frac{V_O^2}{R_L} = \frac{V_I^2 D^2}{2 f_s L}.$$

Rearrangement of this expression yields the dc voltage transfer function of the lossle
buck-boost converter for DCM,

$$M_{V\,DC} = \frac{V_O}{V_I} = D\sqrt{\frac{R_L}{2 f_s L}},$$

which gives

$$D = M_{V\,DC}\sqrt{\frac{2 f_s L}{R_L}}.$$

For the lossy converter,

$$P_O = \eta P_I,$$

which produces

$$\frac{V_O^2}{R_L} = \frac{\eta V_I^2 D^2}{2 f_s L}.$$

Thus,

$$M_{V\,DC} = \frac{V_O}{V_I} = D\sqrt{\frac{\eta R_L}{2 f_s L}}$$

and

$$D - M_{V\,DC}\sqrt{\frac{2 f_s L}{\eta R_L}}.$$

Chapter 5

Flyback PWM DC–DC Converter

1 The dc input voltage of a flyback PWM converter operating in CCM is the US single-phase rectified voltage and the dc output voltage is $V_O = 800\,\text{V}$. Find the transformer turns ratio n.

The minimum dc input voltage occurs at a low line voltage of 90 Vrms and is given by

$$V_{Imin} = \sqrt{2} \times 90 = 127\,\text{V},$$

resulting in the maximum dc voltage transfer function

$$M_{V\,DCmax} = \frac{V_O}{V_{Imin}} = \frac{800}{127} = 6.299.$$

The dc voltage transfer function is

$$M_{V\,DC} = \frac{\eta D}{n(1 - D)}.$$

Assuming the converter efficiency $\eta = 95\%$ and the maximum duty cycle $D_{max} = 0.7$, one obtains

$$n = \frac{\eta D_{max}}{(1 - D_{max})M_{V\,DCmax}} = \frac{0.95 \times 0.7}{(1 - 0.7) \times 6.299} = 0.3519 = \frac{1}{2.842}.$$

Pick $n = 1/3$, that is, the transformer turns ratio is $1:3$. In this case, the maximum duty cycle is

$$D_{max} = \frac{nM_{V\,DCmax}}{nM_{V\,DCmax} + \eta} = \frac{\frac{1}{3} \times 6.299}{\frac{1}{3} \times 6.299 + 0.95} = 0.6885.$$

2 The dc input voltage of a flyback PWM converter is the US single-phase rectified voltage, the dc output voltage is $V_O = 800\,\text{V}$, and the transformer turns ratio $n = 1/3$. Find the voltage stresses of the switch and the diode.

The maximum dc input voltage occurs at a high line voltage of 132 Vrms and is given by

$$V_{Imax} = \sqrt{2} \times 132 = 187\,\text{V}.$$

Hence, the maximum voltage across the switch is

$$V_{SMmax} = V_{Imax} + nV_O = 187 + \frac{1}{3} \times 800 = 454\,\text{V},$$

and the maximum voltage across the diode is

$$V_{DMmax} = \frac{V_{Imax}}{n} + V_O = \frac{187}{\frac{1}{3}} + 800 = 1361\,\text{V}.$$

5.3 The dc input voltage of a flyback PWM converter is the US single-phase rectified voltage, the dc output voltage is $V_O = 800\,\text{V}$, the minimum load current is $I_{Omin} = 0.2$, the maximum input voltage $V_{Imax} = 187\,\text{V}$, the switching frequency is $f_s = 50\,\text{kHz}$, the efficiency is $\eta = 0.95$, and the transformer turns ratio is $n = 1/3$. Find the minimum magnetizing inductance of the transformer to maintain the operation in CCM.

The minimum dc voltage transfer function is

$$M_{V\,DCmin} = \frac{V_O}{V_{Imax}} = \frac{800}{187} = 4.278.$$

Thus, the minimum duty cycle is

$$D_{min} = \frac{nM_{V\,DCmin}}{nM_{V\,DCmin} + \eta} = \frac{\frac{1}{3} \times 4.278}{\frac{1}{3} \times 4.278 + 0.95} = 0.6.$$

The maximum load resistance is

$$R_{Lmax} = \frac{V_O}{I_{Omin}} = \frac{800}{0.2} = 4\,\text{k}\Omega.$$

Hence, the minimum transformer magnetizing inductance is

$$L_{m(min)} = \frac{n^2 R_{Lmax}(1 - D_{min})^2}{2f_s}$$

$$= \frac{\frac{1}{3^2} \times 4 \times 10^3 \times (1 - 0.6)^2}{2 \times 50 \times 10^3} = 0.711\,\text{mH}.$$

Pick $L_m = 2\,\text{mH}$.

5.4 A flyback PWM converter is supplied by the US single-phase rectified line voltage $V_O = 800\,\text{V}$, $I_O = 0.2$ to $0.5\,\text{A}$, the minimum input voltage $V_{Imin} = 127\,\text{V}$, the minimum duty cycle $D_{min} = 0.6$, $n = 1/3$, $L_m = 2\,\text{mH}$, and $f_s = 50\,\text{kHz}$. Find the current stress of the switch and the diode.

The maximum peak-to-peak ripple current of the magnetizing inductance is

$$\Delta i_{Lm(max)} = \frac{nV_O(1 - D_{min})}{f_s L_m} = \frac{\frac{1}{3} \times 800 \times (1 - 0.6)}{50 \times 10^3 \times 2 \times 10^{-3}} = 1.067\,\text{A}.$$

The maximum dc voltage transfer function is

$$M_{V \, DCmax} = \frac{V_O}{V_{Imin}} = \frac{800}{127} = 6.299.$$

The maximum dc input current is

$$I_{Imax} = M_{V \, DCmax} I_{Omax} = 6.299 \times 0.5 = 3.15 \, \text{A}.$$

Hence, the switch current stress is

$$I_{SMmax} = I_{Imax} + \frac{I_{Omax}}{n} + \frac{\Delta i_{Lm(max)}}{2}$$

$$= 3.15 + \frac{0.5}{\frac{1}{3}} + \frac{1.067}{2} = 5.183 \, \text{A}$$

and the diode current stress is

$$I_{DMmax} = nI_{Imax} + I_{Omax} + \frac{n \, \Delta i_{Lm(max)}}{2}$$

$$= \frac{1}{3} \times 3.15 + 0.5 + \frac{\frac{1}{3} \times 1.067}{2} = 1.728 \, \text{A}.$$

5 The dc input voltage of a flyback PWM converter is the US single-phase rectified voltage, the output voltage is $V_O = 800 \, \text{V}$, the load current is $I_O = 0.2$ to $0.5 \, \text{A}$, the maximum dc input current $I_{Imax} = 3.15 \, \text{A}$, the dc voltage transfer function $M_{V \, DCmax} = 6.299$, the transformer turns ratio $n = 1/3$, the switching frequency is $f_s = 50 \, \text{kHz}$, and $V_r/V_O \leq 1 \, \%$. Find the filter capacitance.

The minimum load resistance is

$$R_{Lmin} = \frac{V_O}{I_{Omax}} = \frac{800}{0.5} = 1.6 \, \text{k}\Omega.$$

The ripple voltage is

$$V_r = \frac{V_O}{100} = \frac{800}{100} = 8 \, \text{V}.$$

Let us split this voltage so that the voltage across the capacitance is $V_{Cpp} = 5 \, \text{V}$ and the voltage across the ESR is $V_{rcpp} = 3 \, \text{V}$. The diode current stress is

$$I_{DMmax} = nI_{Imax} + I_{Omax} + \frac{n \, \Delta i_{Lm(max)}}{2}$$

$$= \frac{1}{3} \times 3.15 + 0.5 + \frac{\frac{1}{3} \times 1.067}{2} = 1.728 \, \text{A}.$$

Hence, the maximum resistance of the ESR is

$$r_{Cmax} = \frac{V_{rcpp}}{I_{DMmax}} = \frac{3}{1.728} = 1.736 \, \Omega.$$

Pick $n = 1/3$, that is, the transformer turns ratio is $1:3$. In this case, the maximum duty cycle is

$$D_{max} = \frac{nM_{V\,DCmax}}{nM_{V\,DCmax} + \eta} = \frac{\frac{1}{3} \times 6.299}{\frac{1}{3} \times 6.299 + 0.95} = 0.6885.$$

The minimum filter capacitance

$$C_{min} = \frac{D_{max} V_O}{f_s R_{Lmin} V_{Cpp}} = \frac{0.6885 \times 800}{50 \times 10^3 \times 1600 \times 5} = 1.3768\,\mu\text{F}.$$

Pick $C = 1.5\,\mu\text{F}/1\,\text{kV}/1.5\,\Omega$.

5.7 The dc input voltage of a flyback PWM converter is the US single-phase rectified voltage, the dc output voltage is $V_O = 800\,\text{V}$, the minimum load current is $I_{Omin} = 0\,\text{A}$, the maximum load current is $I_{Omax} = 0.5\,\text{A}$, the switching frequency is $f_s = 50\,\text{kHz}$, the efficiency is $\eta = 0.95$, and the transformer turns ratio is $n = 1/3$. Find the maximum magnetizing inductance of the transformer to maintain the operation in DCM.

The minimum dc input voltage occurs at a low line voltage of 90 Vrms and is given by

$$V_{Imin} = \sqrt{2} \times 90 = 127\,\text{V},$$

resulting in the maximum dc voltage transfer function

$$M_{V\,DCmax} = \frac{V_O}{V_{Imin}} = \frac{800}{127} = 6.299.$$

The dc voltage transfer function is

$$M_{V\,DC} = \frac{\eta D}{n(1 - D)}.$$

The minimum load resistance is

$$R_{Lmin} = \frac{V_O}{I_{Omax}} = \frac{800}{0.5} = 1600\,\Omega.$$

(a) Assuming the converter efficiency $\eta = 95\,\%$ and the maximum duty cycle $D_{max} = 0.7$, one obtains

$$n = \frac{\eta D_{max}}{(1 - D_{max})M_{V\,DCmax}} = \frac{0.95 \times 0.7}{(1 - 0.7) \times 6.299} = 0.3519 = \frac{1}{2.842}.$$

Pick $n = 1/3$, that is, the transformer turns ratio is $1:3$. In this case, the maximum duty cycle is

$$D_{max} = \frac{nM_{V\,DCmax}}{nM_{V\,DCmax} + \eta} = \frac{\frac{1}{3} \times 6.299}{\frac{1}{3} \times 6.299 + 0.95} = 0.6885.$$

The maximum magnetizing inductance is

$$L_{m(max)} = \frac{n^2 R_{Lmin}(1 - D_{Bmax})^2}{2f_s} = \frac{\left(\frac{1}{3}\right)^2 \times 1600(1 - 0.6885)^2}{2 \times 50 \times 10^3} = 172.5\,\mu\text{H}.$$

Pick $L_m = 150\,\mu\text{H}$.

(b) Assuming the converter efficiency $\eta = 95\%$ and the maximum duty cycle $D_{max} = 0.45$, one obtains

$$n = \frac{\eta D_{Bmax}}{(1 - D_{max})M_{V\,DCmax}} = \frac{0.95 \times 0.45}{(1 - 0.45) \times 6.299} = 0.1234 = \frac{1}{8.104}.$$

Pick $n = 1/10$, that is, the transformer turns ratio is $1:10$. In this case, the maximum duty cycle is

$$D_{Bmax} = \frac{nM_{V\,DCmax}}{nM_{V\,DCmax} + \eta} = \frac{\frac{1}{10} \times 6.299}{\frac{1}{10} \times 6.299 + 0.95} = 0.3987.$$

The maximum magnetizing inductance is

$$L_{m(max)} = \frac{n^2 R_{Lmin}(1 - D_{Bmax})^2}{2f_s} = \frac{\left(\frac{1}{10}\right)^2 \times 1600(1 - 0.3987)^2}{2 \times 50 \times 10^3} = 57.849\,\mu H.$$

Pick $L_m = 50\,\mu H$.

Chapter 6
Forward PWM DC–DC Converter

.1 For the forward PWM converter, find the maximum duty cycle D_{MAX}: (a) for $n_3 = 0.5n_1$; (b) for $n_3 = 2n_1$; (c) for $n_3 = 4n_1$.

(a) For $n_3 = 0.5n_1$,

$$D_{MAX} = \frac{1}{\dfrac{n_3}{n_1} + 1} = \frac{1}{\dfrac{0.5n_1}{n_1} + 1} = \frac{1}{0.5 + 1} = 0.6667.$$

(b) For $n_3 = 2n_1$,

$$D_{MAX} = \frac{1}{\dfrac{n_3}{n_1} + 1} = \frac{1}{\dfrac{2n_1}{n_1} + 1} = \frac{1}{2 + 1} = 0.3333.$$

(c) For $n_3 = 4n_1$,

$$D_{MAX} = \frac{1}{\dfrac{n_3}{n_1} + 1} = \frac{1}{\dfrac{4n_1}{n_1} + 1} = \frac{1}{4 + 1} = 0.2.$$

.2 For the forward PWM converter, find the switch peak voltage in terms of the dc input voltage V_I: (a) for $n_3 = 0.5n_1$; (b) for $n_3 = 2n_1$; (c) for $n_3 = 4n_1$.

(a) For $n_3 = 0.5n_1$,

$$V_{SM} = \left(\frac{n_1}{n_3} + 1 \right) V_I = \left(\frac{n_1}{0.5n_1} + 1 \right) V_I = (2 + 1)V_I = 3V_I.$$

se-width Modulated DC–DC Converters – Solutions Manual   Marian K. Kazimierczuk
2008 John Wiley & Sons, Ltd

(b) For $n_3 = 2n_1$,

$$V_{SM} = \left(\frac{n_1}{n_3} + 1\right)V_I = \left(\frac{n_1}{2n_1} + 1\right)V_I = (0.5 + 1)V_I = 1.5V_I.$$

(c) For $n_3 = 4n_1$,

$$V_{SM} = \left(\frac{n_1}{n_3} + 1\right)V_I = \left(\frac{n_1}{4n_1} + 1\right)V_I = (0.25 + 1)V_I = 1.25V_I.$$

6.3 For the forward PWM converter, find the peak voltage of the diode D_3 connected series with the tertiary in terms of the dc input voltage V_I: (a) for $n_3 = 0.5n_1$; (b) $n_3 = 2n_1$; (c) for $n_3 = 4n_1$.

(a) For $n_3 = 0.5n_1$,

$$V_{D3M} = \left(\frac{n_3}{n_1} + 1\right)V_I = \left(\frac{0.5n_1}{n_1} + 1\right)V_I = (0.5 + 1)V_I = 1.5V_I.$$

(b) For $n_3 = 2n_1$,

$$V_{D3M} = \left(\frac{n_3}{n_1} + 1\right)V_I = \left(\frac{2n_1}{n_1} + 1\right)V_I = (2 + 1)V_I = 3V_I.$$

(c) For $n_3 = 4n_1$,

$$V_{D3M} = \left(\frac{n_3}{n_1} + 1\right)V_I = (4 + 1)V_I = 5V_I.$$

6.4 A forward PWM converter is supplied by a European single-phase rectified utility line voltage and $V_O = 12$ V. Find the primary-to-secondary transformer turns ratio n_1.

The minimum dc input voltage is

$$V_{Imin} = \sqrt{2} \times (220 - 0.1 \times 220) = \sqrt{2} \times (220 - 22) = \sqrt{2} \times 198 = 280 \text{ V}.$$

Hence, the maximum dc voltage transfer function is

$$M_{V\,DCmax} = \frac{V_O}{V_{Imin}} = \frac{12}{280} = 0.04286 = \frac{1}{23.3333}.$$

Let us assume the converter efficiency $\eta = 90\%$ and $D_{max} = 0.4$. Thus,

$$n_1 = \frac{\eta D_{max}}{M_{V\,DCmax}} = \frac{0.9 \times 0.4}{0.04286} = 8.3994.$$

Pick $n_1 = 8$.

6.5 A forward PWM converter is supplied by a European single-phase rectified utility line voltage, $V_O = 12$ V, and $n_1 = 8$. Find the maximum permissible duty cycle D_{MAX}.

Pick $n_3 = n_1 = 8$. Hence, the maximum permissible duty cycle D_{MAX} is

$$D_{MAX} = \frac{1}{\dfrac{n_3}{n_1} + 1} = \frac{1}{1 + 1} = 0.5.$$

.6 A forward PWM converter is supplied by an European single-phase rectified utility line voltage, $V_O = 12\,\text{V}$, and $n_1 - n_3 - 8$. Find the minimum, nominal, and maximum duty cycle.

The minimum dc input voltage is

$$V_{Imin} = \sqrt{2} \times (220 - 0.1 \times 220) = \sqrt{2} \times (220 - 22) = \sqrt{2} \times 198 = 280\,\text{V}.$$

The nominal dc input voltage is

$$V_{Inom} = \sqrt{2} \times 220 = 311\,\text{V}.$$

The maximum dc input voltage is

$$V_{Imax} = \sqrt{2} \times (220 + 0.1 \times 220) = \sqrt{2} \times (220 + 22) = \sqrt{2} \times 242 = 342\,\text{V}.$$

Thus, the nominal, minimum, and maximum values of the dc voltage function are

$$M_{V\,DCnom} = \frac{V_O}{V_{Inom}} = \frac{12}{311} = 0.03858 = \frac{1}{25.9167},$$

$$M_{V\,DCmin} = \frac{V_O}{V_{Imax}} = \frac{12}{342} = 0.03509 = \frac{1}{28.5},$$

$$M_{V\,DCmax} = \frac{V_O}{V_{Imin}} = \frac{12}{280} = 0.04286 = \frac{1}{23.3333}.$$

The minimum, nominal, and maximum values of the duty cycle are

$$D_{min} = \frac{n_1 M_{V\,DCmin}}{\eta} = \frac{8 \times 0.03509}{0.9} = 0.3119,$$

$$D_{nom} = \frac{n_1 M_{V\,DCnom}}{\eta} = \frac{8 \times 0.03856}{0.9} = 0.3428,$$

and

$$D_{max} = \frac{n_1 M_{V\,DCmax}}{\eta} = \frac{8 \times 0.04286}{0.9} = 0.381.$$

.7 A forward PWM converter is supplied by a European single-phase rectified utility line voltage, $V_{Imax} = 342\,\text{V}$, $V_O = 12\,\text{V}$, and $n_1 = n_3 = 8$. Find the voltage stresses of the semiconductor devices.

The voltage stress of the switch is

$$V_{SMmax} = \left(\frac{n_1}{n_3} + 1\right) V_{Imax} = 2V_{Imax} = 2 \times 342 = 684\,\text{V}.$$

The voltage stress of the diode D_1 is

$$V_{D1Mmax} = \frac{V_{Imax}}{n_3} = \frac{342}{8} = 42.75\,\text{V}.$$

The voltage stress of the diode D_2 is

$$V_{D2Mmax} = \frac{V_{Imax}}{n_1} = \frac{342}{8} = 42.75\,\text{V}.$$

The voltage stress of the diode D_3 is

$$V_{D3Mmax} = \left(\frac{n_3}{n_1} + 1\right) V_{Imax} = 2V_{Imax} = 2 \times 342 = 684\,\text{V}.$$

6.8   A forward PWM converter has $V_O = 12\,\text{V}$, $I_O = 4$ to $40\,\text{A}$, $n_1 = n_3 = 8$, $V_{Imax} = 342$, $\eta = 90\,\%$, and $f_s = 75\,\text{kHz}$. Find the minimum inductance required for the CCM oper ation.

The minimum dc voltage transfer function is

$$M_{V\,DCmin} = \frac{V_O}{V_{Imax}} = \frac{12}{342} = 0.03509 = \frac{1}{28.5},$$

and the minimum duty cycle is

$$D_{min} = \frac{n_1 M_{V\,DCmin}}{\eta} = \frac{8 \times 0.03509}{0.9} = 0.3119.$$

The maximum resistance is

$$R_{Lmax} = \frac{V_O}{I_{Omin}} = \frac{12}{4} = 3\,\Omega.$$

The minimum inductance is

$$L_{min} = \frac{R_{Lmax}(1 - D_{min})}{2f_s} = \frac{3 \times (1 - 0.3119)}{2 \times 75 \times 10^3} = 13.762\,\mu\text{H}.$$

Pick $L = 20\,\mu\text{H}$.

6.9   A forward PWM converter has $V_O = 12\,\text{V}$, $I_O = 4$ to $40\,\text{A}$, duty cycle $D_{min} = 0.31$, $L = 20\,\mu\text{H}$, $f_s = 75\,\text{kHz}$, and $V_r/V_O \leq 1\,\%$. Find the filter capacitance and the corr frequency of the output filter.

The maximum peak-to-peak value of the inductor current ripple is

$$\Delta i_{Lmax} = \frac{V_O(1 - D_{min})}{f_s L} = \frac{12 \times (1 - 0.3119)}{75 \times 10^3 \times 20 \times 10^{-6}} = 5.505\,\text{A}.$$

The ripple voltage is

$$V_r = \frac{V_O}{100} = \frac{12}{100} = 120 \, \text{mV}.$$

Hence, the maximum ESR is

$$r_{Cmax} = \frac{V_r}{\Delta i_{Lmax}} = \frac{0.12}{5.505} = 21.8 \, \text{m}\Omega.$$

Pick $r_C = 20 \, \text{m}\Omega$. Therefore, the minimum filter capacitance is

$$C_{min} = \max\left\{\frac{D_{max}}{2f_s r_C}, \frac{1 - D_{min}}{2f_s r_C}\right\} = \max\left\{\frac{0.381}{2f_s r_C}, \frac{1 - 0.3119}{2f_s r_C}\right\}$$

$$= \frac{1 - D_{min}}{2f_s r_C} = \frac{1 - 0.3119}{2 \times 75 \times 10^3 \times 0.02} = 229 \, \mu\text{F}.$$

Let $C = 300 \, \mu\text{F}/25 \, \text{V}/0.02 \, \Omega$.

The corner frequency of the output low-pass filter is

$$f_o = \frac{1}{2\pi\sqrt{LC}} = \frac{1}{2\pi\sqrt{20 \times 10^{-6} \times 300 \times 10^{-6}}} = 2.055 \, \text{kHz}.$$

10 A forward PWM converter has $V_O = 12 \, \text{V}$, $I_O = 4$ to $40 \, \text{A}$, $L = 20 \, \mu\text{H}$, $D_{min} = 0.3119$, and $f_s = 75 \, \text{kHz}$. Find the current stresses of the rectifier diodes.

The maximum peak-to-peak value of the inductor current ripple is

$$\Delta i_{Lmax} = \frac{V_O(1 - D_{min})}{f_s L} = \frac{12 \times (1 - 0.3119)}{75 \times 10^3 \times 20 \times 10^{-6}} = 5.505 \, \text{A}.$$

The current stresses of the rectifier diodes D_1 and D_2 are

$$I_{D1Mmax} = I_{D2Mmax} = I_{Omax} + \frac{\Delta i_{Lmax}}{2}$$

$$= 40 + \frac{5.505}{2} = 42.753 \, \text{A}.$$

11 A forward PWM converter has $V_O = 12 \, \text{V}$, $I_O = 4$ to $40 \, \text{A}$, $n_1 = n_3 = 8$, $L = 20 \, \mu\text{H}$, $V_{Imax} = 342 \, \text{V}$, $D_{min} = 0.3119$, and $f_s = 75 \, \text{kHz}$. Find the magnetizing inductance such that its peak current is less than 12 % of the maximum peak current of the primary of the ideal transformer.

From Problem 6.10, the current stress of diode D_1 is $I_{D1Mmax} = 42.753 \, \text{A}$. Thus, the maximum peak current of the diode D_1 reflected to the primary winding of the ideal transformer is

$$I_{1max} = \frac{I_{D1Mmax}}{n_1} = \frac{42.753}{8} = 5.344 \, \text{A}.$$

Hence, the maximum peak current through the magnetizing inductance is

$$\Delta i_{Lm(max)} = 0.12 I_{1max} = 0.12 \times 5.344 = 0.641 \, \text{A},$$

resulting in the minimum value of the magnetizing inductance

$$L_{m(min)} = \frac{D_{min} V_{Imax}}{f_s \Delta i_{Lm(max)}} = \frac{0.3119 \times 342}{75 \times 10^3 \times 0.641} = 2.219 \, \text{mH}.$$

Pick $L_m = 2.3 \, \text{mH}$.

6.12 A forward PWM converter is supplied by a European single-phase rectified utility li■ voltage, $V_O = 12 \, \text{V}$, $I_O = 4$ to $40 \, \text{A}$, $n_1 = n_3 = 8$, $L = 20 \, \mu\text{H}$, and $f_s = 75 \, \text{kHz}$. Fi■ the current stress of the switch.

The maximum peak current through the ideal transformer primary winding found Problem 6.11 is $I_{1max} = 5.344 \, \text{A}$ and the maximum peak current through the magneti■ ing inductance $\Delta i_{Lm(max)} = 0.641 \, \text{A}$. Hence, the current stress of the switch is

$$I_{SMmax} = I_{1max} + \Delta i_{Lm(max)} = 5.344 + 0.641 = 5.985 \, \text{A}.$$

6.13 For a forward PWM converter supplied by a European single-phase rectified utility li■ voltage, determine the maximum inductance required for DCM operation, if $V_O = 12$ ˈ $I_O = 0$ to $4 \, \text{A}$, and $f_s = 75 \, \text{kHz}$.

The minimum dc input voltage is

$$V_{Imin} = \sqrt{2} \times (220 - 0.1 \times 220) = \sqrt{2} \times (220 - 22) = \sqrt{2} \times 198 = 280 \, \text{V}.$$

Hence, the maximum dc voltage transfer function is

$$M_{V \, DCmax} = \frac{V_O}{V_{Imin}} = \frac{12}{280} = 0.04286 = \frac{1}{23.3333}.$$

Assuming $\eta = 0.9$, the maximum duty cycle at the CCM/DCM boundary is

$$D_{Bmax} = \frac{n_1 M_{V \, DCmax}}{\eta} = \frac{8 \times 0.04286}{0.9} = 0.381.$$

The minimum resistance is

$$R_{Lmin} = \frac{V_O}{I_{Omax}} = \frac{12}{4} = 3 \, \Omega.$$

The maximum inductance required for DCM is

$$L_{max} = \frac{R_{Lmin}(1 - D_{Bmax})}{2 f_s} = \frac{3 \times (1 - 0.381)}{2 \times 75 \times 10^3} = 12.38 \, \mu\text{H}.$$

Chapter 7

Half-bridge PWM DC–DC Converter

1 Derive an expression for the voltage stress of the diodes in the half-bridge PWM converter with a bridge rectifier.

The voltage across the primary winding is

$$v_1 = \frac{V_I}{2},$$

and the voltage across the secondary is

$$v_2 = \frac{v_1}{n} = \frac{V_I}{2n}.$$

The voltages across the diodes in the off state are

$$v_{D1} = v_{D2} = v_{D3} = v_{D4} = -v_2 = -\frac{v_1}{n} = -\frac{V_I}{2n}.$$

Hence, the voltage stress of the diodes is

$$V_{DMmax} = \frac{V_{Imax}}{2n}.$$

2 The input voltage of a half-bridge PWM converter operating in CCM is the European single-phase rectified line voltage 220 Vrms ± 10 % and $V_O = 12$ V. Find the transformer turns ratio n, the minimum duty cycle D_{min}, and the maximum duty cycle D_{max}.

The dc minimum and maximum voltages are

$$V_{Imin} = \sqrt{2} \times (220 - 0.10 \times 220) = \sqrt{2} \times (220 - 22) = \sqrt{2} \times 198 = 280 \, \text{V}$$

and

$$V_{Imax} = \sqrt{2} \times (220 + 0.1 \times 220) = \sqrt{2} \times (220 + 22) = \sqrt{2} \times 242 = 342 \, \text{V}.$$

The minimum and maximum dc voltage transfer functions are

$$M_{V\,DCmin} = \frac{V_O}{V_{Imax}} = \frac{12}{342} = 0.035 = \frac{1}{28.5}$$

and

$$M_{V\,DCmax} = \frac{V_O}{V_{Imin}} = \frac{12}{280} = 0.04286 = \frac{1}{23.3333}.$$

Assuming the converter efficiency $\eta = 92\,\%$ and the maximum duty cycle $D_{max} = 0.4$ we have

$$n = \frac{\eta D_{max}}{M_{V\,DCmax}} = \frac{0.92 \times 0.42}{0.04286} = 9.015.$$

Let $n = 9$. The minimum and maximum values of the duty cycle are

$$D_{min} = \frac{nM_{V\,DCmin}}{\eta} = \frac{9 \times 0.035}{0.92} = 0.3424$$

and

$$D_{max} = \frac{nM_{V\,DCmax}}{\eta} = \frac{9 \times 0.04286}{0.92} = 0.4193.$$

7.3 A half-bridge converter operating in CCM has $V_{Imax} = 342\,$V and $n = 9$. Find the voltage stresses of the switches and the diodes.

The voltage stress of the switches is

$$V_{SMmax} = V_{Imax} = 342\,\text{V}.$$

If a transformer center-tapped rectifier is used, the voltage stress of the diodes is

$$V_{DMmax} = \frac{V_{Imax}}{n} = \frac{342}{9} = 38\,\text{V}.$$

7.4 A half-bridge converter has $V_{Imin} = 280\,$V, $V_{Imax} = 342\,$V, $V_O = 12\,$V, $I_O = 1$ to 20, $f_s = 50\,$kHz and $D_{min} = 0.3424$. Find the minimum inductance required to maintain the converter operation in CCM.

The minimum and maximum load resistances are

$$R_{Lmin} = \frac{V_O}{I_{Omax}} = \frac{12}{20} = 0.6\,\Omega$$

and

$$R_{Lmax} = \frac{V_O}{I_{Omin}} = \frac{12}{1} = 12\,\Omega.$$

The minimum inductance is

$$L_{min} = \frac{R_{Lmax}(0.5 - D_{min})}{2f_s} = \frac{12 \times (0.5 - 0.3424)}{2 \times 50 \times 10^3} = 18.9\,\mu\text{H}.$$

Pick $L = 25\,\mu$H.

5 A half-bridge converter has $V_{Imin} = 280\,\text{V}$, $V_{Imax} = 342\,\text{V}$, $V_O = 12\,\text{V}$, $I_O = 1$ to $20\,\text{A}$, $L = 25\,\mu\text{H}$, $n = 9$, $D_{min} = 0.3424$, $D_{max} = 0.4193$, $f_s = 50\,\text{kHz}$, and $V_r/V_O < 2\,\%$. Find the minimum filter capacitance and the corner frequency of the output filter.

The maximum peak-to-peak value of the inductor ripple current is

$$\Delta i_{Lmax} = \frac{V_O(0.5 - D_{min})}{f_s L} = \frac{12 \times (0.5 - 0.3424)}{50 \times 10^3 \times 25 \times 10^{-6}} = 1.513\,\text{A}.$$

The ripple voltage is

$$V_r = \frac{V_O}{50} = \frac{12}{50} = 0.24\,\text{V}.$$

Hence, the maximum ESR is

$$r_{Cmax} = \frac{V_r}{\Delta i_{Lmax}} = \frac{240 \times 10^{-3}}{1.513} = 158.6\,\text{m}\Omega.$$

Pick $r_C = 0.1\,\Omega$. The minimum capacitance is

$$C_{min} = \max \left\{ \frac{D_{max}}{2 f_s r_C}, \frac{0.5 - D_{min}}{2 f_s r_C} \right\} = \max \left\{ \frac{0.4193}{2 f_s r_C}, \frac{0.5 - 0.3424}{2 f_s r_C} \right\}$$

$$= \frac{D_{max}}{2 f_s r_C} = \frac{0.4193}{2 \times 50 \times 10^3 \times 0.1} = 41.93\,\mu\text{F}.$$

Pick $C = 50\,\mu\text{F}/25\,\text{V}/0.1\,\Omega$. The corner frequency of the output filter is then

$$f_o = \frac{1}{2\pi \sqrt{LC}} = \frac{1}{2\pi \sqrt{25 \times 10^{-6} \times 50 \times 10^{-6}}} = 4.5\,\text{kHz}.$$

6 A half-bridge converter has $V_{Imin} = 280\,\text{V}$, $V_{Imax} = 342\,\text{V}$, $V_O = 12\,\text{V}$, $I_O = 1$ to $20\,\text{A}$, $L = 25\,\mu\text{H}$, $n = 9$, $\Delta i_{Lmax} = 1.513$, and $f_s = 50\,\text{kHz}$. Determine the minimum magnetizing inductance at which its peak-to-peak current is less than $10\,\%$ of the maximum peak current of the ideal transformer primary.

The maximum peak current of the primary of the ideal transformer is

$$I_{1max} = \frac{I_{Omax}}{n} + \frac{\Delta i_{Lmax}}{2n} = \frac{20}{9} + \frac{1.513}{2 \times 9} = 2.222 + 0.084 = 2.306\,\text{A}.$$

The maximum peak-to-peak value of the magnetizing inductance current is

$$\Delta i_{Lm(max)} = 0.1 I_{1max} = 0.1 \times 2.306 = 0.2306\,\text{A}.$$

Thus, the minimum magnetizing inductance is

$$L_{m(min)} = \frac{D_{min} V_{Imax}}{2 f_s \Delta i_{Lm(max)}} = \frac{0.3424 \times 342}{2 \times 50 \times 10^3 \times 0.2306} = 5.08\,\text{mH}.$$

Pick $L_m = 5.2\,\text{mH}$.

7.7 A half-bridge converter has $V_{Imin} = 280$ V, $V_{Imax} = 342$ V, $V_O = 12$ V, $I_O = 0$ to 20.
$n = 9, f_s = 50$ kHz, and $D_{min} = 0.3424$. Find the maximum inductance required to maintain operation of the converter in DCM.

The minimum and maximum load resistances are

$$R_{Lmin} = \frac{V_O}{I_{Omax}} = \frac{12}{20} = 0.6\,\Omega.$$

The maximum dc voltage transfer function is

$$M_{V\,DCmax} = \frac{V_O}{V_{Imin}} = \frac{12}{280} = 0.04286 = \frac{1}{23.3333}$$

The minimum duty cycle at the boundary is

$$D_{Bmax} = nM_{V\,DCmax} = 9 \times 0.04286 = 0.3857.$$

Thus, the maximum inductance is

$$L_{max} = \frac{R_{Lmin}(0.5 - D_{Bmax})}{2f_s} = \frac{0.6 \times (0.5 - 0.3857)}{2 \times 50 \times 10^3} = 0.6858\,\mu\text{H}.$$

Pick $L = 0.6\,\mu$H.

Chapter 8

Full-bridge PWM DC–DC Converter

.1 Derive an expression for the voltage stress of the diodes in the full-bridge converter with a full-bridge rectifier.

The voltage across the primary winding is

$$v_1 = V_I,$$

and the voltage across the secondary winding is

$$v_2 = \frac{v_1}{n} = \frac{V_I}{n}.$$

The voltages across the diodes in the off state are

$$v_{D1} = v_{D2} = v_{D3} = v_{D4} = -v_2 = -\frac{v_1}{n} = -\frac{V_I}{n}.$$

Hence, the voltage stresses of the diodes are given by

$$V_{DMmax} = \frac{V_{Imax}}{n}.$$

.2 A full-bridge PWM converter has input voltage from a US single-phase rectified line, $V_O = 48$ V, $P_O = 1$ to 2.5 kW, and $f_s = 35$ kHz. Find the transformer turns ratio n, the minimum duty cycle D_{min}, and the maximum duty cycle D_{max}.

The minimum dc input voltage is

$$V_{Imin} = \sqrt{2} \times 90 = 127 \,\text{V},$$

and the maximum dc input voltage is

$$V_{Imax} = \sqrt{2} \times 132 = 187 \,\text{V}.$$

The maximum load current is

$$I_{Omax} = \frac{P_{Omax}}{V_O} = \frac{2500}{48} = 52\,\text{A},$$

and the minimum load current is

$$I_{Omin} = \frac{P_{Omin}}{V_O} = \frac{1000}{48} = 20.833\,\text{A}.$$

The minimum load resistance is

$$R_{Lmin} = \frac{V_O}{I_{Omax}} = \frac{48}{52} = 0.923\,\Omega$$

and the maximum load resistance is

$$R_{Lmax} = \frac{V_O}{I_{Omin}} = \frac{48}{20.833} = 2.304\,\Omega.$$

The minimum and maximum values of the dc voltage transfer function are

$$M_{V\,DCmin} = \frac{V_O}{V_{Imax}} = \frac{48}{187} = 0.2567 = \frac{1}{3.8958}$$

and

$$M_{V\,DCmax} = \frac{V_O}{V_{Imin}} = \frac{48}{127} = 0.378 = \frac{1}{2.6458}.$$

Assume the converter efficiency $\eta = 0.95$ and the maximum duty cycle $D_{max} = 0$
The transformer turns ratio is

$$n = \frac{2\eta D_{max}}{M_{V\,DCmax}} = \frac{2 \times 0.95 \times 0.4}{0.377} = 2.0159.$$

Let $n = 2$. The minimum and maximum values of the duty cycle are

$$D_{min} = \frac{nM_{V\,DCmin}}{2\eta} = \frac{2 \times 0.2567}{2 \times 0.95} = 0.27021$$

and

$$D_{max} = \frac{nM_{V\,DCmax}}{2\eta} = \frac{2 \times 0.377}{2 \times 0.95} = 0.3968.$$

8.3 A full-bridge PWM converter has $V_{Imin} = 127\,\text{V}$, $V_{Imax} = 187\,\text{V}$, $n = 2$, $V_O = 48$
$P_O = 1$ to $2.5\,\text{kW}$, and $f_s = 35\,\text{kHz}$. Find the voltage stresses of the transistors and t
diodes.

The voltage stress across each switch is

$$V_{SMmax} = V_{Imax} = 187\,\text{V}.$$

If a transformer center-tapped rectifier is used, the voltage stress across each diode

$$V_{DMmax} = \frac{2V_{Imax}}{n} = \frac{2 \times 187}{2} = 187\,\text{V}.$$

If a full-bridge rectifier is used, the voltage stress across each diode is

$$V_{DMmax} = \frac{V_{Imax}}{n} = \frac{187}{2} = 93.5\,\text{V}.$$

3.4 A full-bridge PWM converter has $V_{Imin} = 127\,\text{V}$, $V_{Imax} = 187\,\text{V}$, $n = 2$, $V_O = 48\,\text{V}$, $P_O = 1$ to $2.5\,\text{kW}$, $D_{min} = 0.27$, and $f_s = 35\,\text{kHz}$. Find the minimum inductance required to maintain the converter operation in CCM. Calculate the maximum value of peak-to-peak inductor ripple current and $\Delta i_{Lmax}/I_{Omax}$.

The minimum load current is

$$I_{Omin} = \frac{P_{Omin}}{V_O} = \frac{1000}{48} = 20.833\,\text{A},$$

and the maximum load resistance is

$$R_{Lmax} = \frac{V_O}{I_{Omin}} = \frac{48}{20.833} = 2.304\,\Omega.$$

We have $D_{min} = 0.27$. Hence, the minimum inductance is

$$L_{min} = \frac{R_{Lmax}(0.5 - D_{min})}{2f_s} = \frac{2.304 \times (0.5 - 0.27)}{2 \times 35 \times 10^3} = 7.57\,\mu\text{H}.$$

Let $L = 10\,\mu\text{H}$.

The maximum inductor ripple current is

$$\Delta i_{Lmax} = \frac{V_O(0.5 - D_{min})}{f_s L} = \frac{48 \times (0.5 - 0.27)}{35 \times 10^3 \times 10 \times 10^{-6}} = 31.5428\,\text{A}.$$

Hence,

$$\frac{\Delta i_{Lmax}}{I_{Omax}} = \frac{31.5428}{52} = 0.606 = 60.6\%.$$

3.5 A full-bridge PWM converter has $V_{Imin} = 127\,\text{V}$, $V_{Imax} = 187\,\text{V}$, $n = 2$, $V_O = 48\,\text{V}$, $P_O = 1$ to $2.5\,\text{kW}$, $D_{min} = 0.27$, and $f_s = 35\,\text{kHz}$. Find the minimum inductance at which the ratio of the peak-to-peak inductor ripple current to the maximum load current is less than $10\,\%$.

The maximum value of the peak-to-peak inductor ripple current in terms of the maximum load current is

$$\Delta i_{Lmax} = \frac{V_O(0.5 - D_{min})}{f_s L_{min}} = \frac{I_{Omax} R_{Lmin}(0.5 - D_{min})}{f_s L_{min}}.$$

Hence,

$$\frac{\Delta i_{Lmax}}{I_{Omax}} = \frac{R_{Lmin}(0.5 - D_{min})}{f_s L_{min}}.$$

Thus,

$$L_{min} = \frac{R_{Lmin}(0.5 - D_{min})}{f_s} \frac{I_{Omax}}{\Delta i_{Lmax}} = \frac{0.923 \times (0.5 - 0.27)}{35 \times 10^3} \times 10$$

$$= 60.65\,\mu\text{H}.$$

Let $L = 70\,\mu\text{H}$.

8.6　A full-bridge PWM converter has input voltage from a US single-phase rectified line
$V_O = 48\,\text{V}$, $P_O = 1$ to $2.5\,\text{kW}$, $L = 70\,\mu\text{H}$, $f_s = 35\,\text{kHz}$, and $V_r/V_O \le 1\,\%$. Find t
filter capacitance and the corner frequency of the output filter.

The maximum inductor ripple current is

$$\Delta i_{Lmax} = \frac{V_O(0.5 - D_{min})}{f_s L} = \frac{48 \times (0.5 - 0.27)}{35 \times 10^3 \times 70 \times 10^{-6}} = 4.506\,\text{A}.$$

The ripple voltage is

$$V_r = \frac{V_O}{100} = \frac{48}{100} = 0.48\,\text{V}.$$

The maximum filter capacitor ESR is

$$r_{Cmax} = \frac{V_r}{\Delta i_{Lmax}} = \frac{0.48}{4.506} = 106.52\,\text{m}\Omega.$$

Pick $r_C = 75\,\text{m}\Omega$. Hence, the minimum filter capacitance is

$$C_{min} = \max\left\{ \frac{D_{max}}{2f_s r_C}, \frac{0.5 - D_{min}}{2f_s r_C} \right\} = \max\left\{ \frac{0.3968}{2f_s r_C}, \frac{0.5 - 0.27}{2f_s r_C} \right\}$$

$$= \frac{D_{max}}{2f_s r_C} = \frac{0.3968}{2 \times 35 \times 10^3 \times 0.075} = 75.58\,\mu\text{F}.$$

Pick $C = 100\,\mu\text{F}/100\,\text{V}/75\,\text{m}\Omega$. The corner frequency of the output filter is

$$f_o = \frac{1}{2\pi\sqrt{LC}} = \frac{1}{2\pi\sqrt{70 \times 10^{-6} \times 100 \times 10^{-6}}} = 1.9\,\text{kHz}.$$

8.7　A full-bridge PWM converter has input voltage from a US single-phase rectified li
$V_O = 48\,\text{V}$, $P_O = 1$ to $2.5\,\text{kW}$, $L = 70\,\mu\text{H}$, and $f_s = 35\,\text{kHz}$, and $V_r/V_O \le 1\,\%$. Find
minimum magnetizing inductance $L_{m(min)}$ at which its maximum peak-to-peak curren
less then $10\,\%$ of the maximum peak current of the ideal transformer primary windin

The maximum peak current of the primary winding of the ideal transformer is

$$I_{1max} = \frac{I_{Omax}}{n} + \frac{\Delta i_{Lmax}}{2n} = \frac{52}{2} + \frac{31.5428}{2 \times 2} = 26 + 7.885 = 33.886\,\text{A}.$$

The maximum peak-to-peak value of the magnetizing inductance current is

$$\Delta i_{Lm(max)} = 0.1 I_{1max} = 0.1 \times 33.885 = 3.388\,\text{A}.$$

Hence,

$$L_{m(min)} = \frac{D_{min} V_{Imax}}{f_s \Delta i_{Lm(max)}} = \frac{0.27 \times 187}{35 \times 10^3 \times 3.388} = 425.788 \, \mu H.$$

8.8 A full-bridge PWM converter has $V_{Imin} = 127 \, V$, $V_{Imax} = 187 \, V$, $n = 2$, $V_O = 48 \, V$, $P_O - 1$ to 2.5 kW, $L = 70 \, \mu F$, and $f_s = 35 \, kHz$. Find the current stresses of the transistors and the diodes.

The switch current stress is

$$I_{SMmax} = \frac{I_{Omax}}{n} + \frac{\Delta i_{Lmax}}{2n} + \frac{\Delta i_{Lm(max)}}{2}$$

$$= \frac{52}{2} + \frac{31.5428}{2 \times 2} + \frac{3.388}{2} = 26 + 7.885 + 1.694 = 35.58 \, A$$

and the diode current stress is

$$I_{DMmax} = I_{Omax} + \frac{\Delta i_{Lmax}}{2} = 52 + \frac{31.5428}{2} = 67.7714 \, A.$$

8.9 A full-bridge dc–dc converter accepts the US single-phase rectified line and delivers $V_O = 1 \, kV$. Find the transformer turns ratio n, the minimum duty cycle D_{min}, and the maximum duty cycle D_{max}.

The minimum dc input voltage of the converter is

$$V_{Imin} = \sqrt{2} \times 90 = 127 \, V,$$

and the maximum dc input voltage is

$$V_{Imax} = \sqrt{2} \times 132 = 187 \, V.$$

The minimum dc voltage function is

$$M_{V \, DCmin} = \frac{V_O}{V_{Imax}} = \frac{1000}{187} = 5.3476,$$

and the maximum dc voltage function is

$$M_{V \, DCmax} = \frac{V_O}{V_{Imin}} = \frac{1000}{127} = 7.874.$$

Assuming the converter efficiency $\eta = 95 \, \%$ and the maximum duty cycle $D_{max} = 0.4$, we have

$$n = \frac{2 \eta D_{max}}{M_{V \, DCmax}} = \frac{2 \times 0.95 \times 0.4}{7.874} = 0.09652 = \frac{1}{10.363}.$$

Pick $n = 1/10$, that is the transformer turns ratio is $1 : 10$. This gives

$$D_{min} = \frac{n M_{V \, DCmin}}{2 \eta} = \frac{0.1 \times 5.3476}{2 \times 0.95} = 0.2815$$

and

$$D_{max} = \frac{nM_{V\,DCmax}}{2\eta} = \frac{0.1 \times 7.874}{2 \times 0.95} = 0.4144.$$

8.10 A full-bridge dc–dc converter with a bridge rectifier accepts the US single-phase rectified line and has $V_O = 1\,\text{kV}$ and $n = 1/10$. Find the voltage stresses of the switches and the diodes.

The voltage stress of the switches is

$$V_{SMmax} = V_{Imax} = \sqrt{2} \times 132 = 187\,\text{V}.$$

Since the output voltage is high, a full-bridge rectifier is used. The voltage stress of the diodes is

$$V_{DMmax} = \frac{V_{Imax}}{n} = \frac{187}{0.1} = 1870\,\text{V}.$$

8.11 A full-bridge dc–dc converter accepts the US single-phase rectified line and has V_O 1 kV, $n = 1/10$, $P_O = 100\,\text{W}$ to 1 kW, $D_{min} = 0.2815$, and $f_s = 50\,\text{kHz}$. Find the minimum inductance for CCM operation.

The minimum load current is

$$I_{Omin} = \frac{P_{Omin}}{V_O} = \frac{100}{1000} = 0.1\,\text{A}.$$

The maximum load resistance is

$$R_{Lmax} = \frac{V_O}{I_{Omin}} = \frac{1000}{0.1} = 10\,\text{k}\Omega.$$

From Problem 8.9, $D_{min} = 0.2815$. Hence,

$$L_{min} = \frac{R_{Lmax}(0.5 - D_{min})}{2f_s} = \frac{10 \times 10^3 \times (0.5 - 0.2815)}{2 \times 50 \times 10^3} = 21.85\,\text{mH}.$$

Pick $L = 25\,\text{mH}$.

8.12 A full-bridge dc–dc converter accepts the US single-phase rectified line and has V_O 1 kV, $n = 1/10$, $P_O = 100\,\text{W}$ to 1 kW, $D_{min} = 0.2814$, $D_{max} = 0.4144$, $L = 25\,\text{mH}$ and $f_s = 50\,\text{kHz}$. Find the minimum filter capacitance and its minimum ESR.

The maximum peak-to-peak inductor current is

$$\Delta i_{Lmax} = \frac{V_O(0.5 - D_{min})}{f_s L} = \frac{1000 \times (0.5 - 0.2815)}{50 \times 10^3 \times 25 \times 10^{-3}} = 0.1748\,\text{A}.$$

The ripple voltage is

$$V_r = 0.01 V_O = 0.01 \times 1000 = 10\,\text{V}.$$

The maximum ESR is

$$r_{Cmax} = \frac{V_r}{\Delta i_{Lmax}} = \frac{10}{0.17488} = 57.21\,\Omega.$$

Pick $r_C = 50\,\Omega$. The minimum filter capacitance is

$$C_{min} = \max\left\{\frac{D_{max}}{2f_s r_C}, \frac{0.5 - D_{min}}{2f_s r_C}\right\} = \max\left\{\frac{0.4144}{2f_s r_C}, \frac{0.5 - 0.2815}{2f_s r_C}\right\}$$

$$= \frac{D_{max}}{2f_s r_C} = \frac{0.4144}{2 \times 50 \times 10^3 \times 50 \times 10^{-3}} = 82.88\,\mu\text{F}.$$

Let $C = 100\,\mu\text{F}/1200\,\text{V}/50\,\Omega$.

Chapter 9

Push-pull PWM DC–DC Converter

1 A push-pull PWM converter has $V_I = 48 \pm 6\,V$ and $V_O = 180\,V$. Find the transformer turns ratio n.

The maximum voltage transfer function is

$$M_{V\,DCmax} = \frac{V_O}{V_{Imin}} = \frac{180}{42} = 4.2857.$$

Assume $\eta = 95\,\%$ and $D_{max} = 0.45$. Thus,

$$n = \frac{2\eta D_{max}}{M_{V\,DCmax}} = \frac{2 \times 0.95 \times 0.45}{4.2857} = 0.1995 = \frac{1}{5.0125}.$$

Pick $n = 1/5$.

2 A push-pull PWM converter has $V_I = 48 \pm 6\,V$, $V_O = 180\,V$, and $n = 0.2$. Find the maximum and minimum duty cycle.

From Problem 9.1, $M_{V\,DCmax} = 4.2857$, so the maximum duty cycle is

$$D_{max} = \frac{nM_{V\,DCmax}}{2\eta} = \frac{\frac{1}{5} \times 4.2857}{2 \times 0.95} = 0.4511.$$

The minimum voltage transfer function is

$$M_{V\,DCmin} = \frac{V_O}{V_{Imax}} = \frac{180}{54} = 3.3333.$$

The minimum duty cycle is

$$D_{min} = \frac{nM_{V\,DCmin}}{2\eta} = \frac{\frac{1}{5} \times 3.3333}{2 \times 0.95} = 0.3509.$$

9.3 A push-pull PWM converter has $V_I = 48 \pm 6\,\text{V}$, $V_O = 180\,\text{V}$, and $n = 0.2$. Find the voltage stresses of the switches and the diodes.

The voltage stress of the switches is

$$V_{SMmax} = 2V_{Imax} = 2 \times 54 = 108\,\text{V}.$$

The output voltage is high; therefore, a transformer full-bridge rectifier will be used. The voltage stress of the diodes is

$$V_{DMmax} = \frac{V_{Imax}}{n} = \frac{54}{0.2} = 270\,\text{V}.$$

9.4 A push-pull PWM converter has $V_I = 48 \pm 6\,\text{V}$, $V_O = 180\,\text{V}$, $n = 0.2$, $I_O = 0.2$ to 2, $D_{min} = 0.3509$, and $f_s = 40\,\text{kHz}$. Find the minimum inductance.

The maximum load resistance is

$$R_{Lmax} = \frac{V_O}{I_{Omin}} = \frac{180}{0.2} = 900\,\Omega.$$

The minimum inductance is

$$L_{min} = \frac{R_{Lmax}(0.5 - D_{min})}{2f_s} = \frac{900 \times (0.5 - 0.3509)}{2 \times 40 \times 10^3} = 1.6785\,\text{mH}.$$

Pick $L_{min} = 2\,\text{mH}$.

9.5 A push-pull PWM converter has $V_I = 48 \pm 6\,\text{V}$, $V_O = 180\,\text{V}$, $n = 0.2$, $I_O = 0.2$ to 2, $L = 2\,\text{mH}$, $D_{min} = 0.3509$, $f_s = 40\,\text{kHz}$, and $V_r/V_O \leq 0.5\%$. Find the minimum filter capacitance, the corner frequency of the output filter, and the ratio of the output ripple frequency to the corner frequency.

$$\Delta i_{Lmax} = \frac{V_O(0.5 - D_{min})}{f_s L_{min}} = \frac{180 \times (0.5 - 0.3509)}{40 \times 10^3 \times 2 \times 10^{-3}} = 0.3357\,\text{A}.$$

The ripple voltage is

$$V_r = 0.005V_O = 0.005 \times 180 = 0.9\,\text{V}.$$

The maximum ESR is

$$r_{Cmax} = \frac{V_r}{\Delta i_{Lmax}} = \frac{0.9}{0.3357} = 2.68\,\Omega.$$

Pick $r_C = 1\,\Omega$.

From Problem 9.2, $D_{max} = 0.4511$. Hence, the minimum filter capacitance is

$$C_{min} = \max\left\{\frac{D_{max}}{2r_C f_s}, \frac{0.5 - D_{min}}{2r_C f_s}\right\} = \max\left\{\frac{0.4511}{2r_C f_s}, \frac{0.5 - 0.3509}{2r_C f_s}\right\}$$

$$= \frac{D_{max}}{2r_C f_s} = \frac{0.4511}{2 \times 1 \times 40 \times 10^3} = 5.639\,\mu\text{F}.$$

Pick $C = 10\,\mu\text{F}/250\,\text{V}/1\,\Omega$.

The corner frequency of the output filter is

$$f_o = \frac{1}{2\pi\sqrt{LC}} = \frac{1}{2\pi\sqrt{2 \times 10^{-3} \times 10 \times 10^{-6}}} = 1.125\,\text{kHz}.$$

Thus, the ratio of the output ripple to the corner frequency is

$$\frac{f_{ripple}}{f_o} = \frac{2f_s}{f_0} = \frac{2 \times 40}{1.125} = 71.11.$$

6 A push-pull PWM converter has $V_I = 48 \pm 6\,\text{V}$, $V_O = 180\,\text{V}$, $n = 0.2$, $I_O = 0.2$ to $2\,\text{A}$, $L = 2\,\text{mH}$, $f_s = 40\,\text{kHz}$, and $V_r/V_O \leq 0.5\,\%$. Find the current stresses of the switches and the diodes.

The current stress of the diodes is

$$I_{DMmax} = I_{Omax} + \frac{\Delta i_{Lmax}}{2} = 2 + \frac{0.3357}{2} = 2.168\,\text{A}.$$

The maximum peak current of the ideal transformer primary winding is

$$\Delta i_{Lm1(max)} = I_{1max} = \frac{I_{DMmax}}{n} = \frac{2.168}{0.2} = 10.84\,\text{A}.$$

Let us assume that the maximum peak-to-peak current of the magnetizing inductance is less than 10 % of the maximum peak current of the ideal transformer primary winding. The current stress of the switches is

$$I_{SMmax} = \frac{I_{Omax}}{n} + \frac{\Delta i_{Lmax}}{2n} + \frac{\Delta i_{Lm1(max)}}{2} = \frac{2}{0.2} + \frac{0.3357}{2 \times 0.2} + \frac{10.84}{2}$$
$$= 16.26\,\text{A}.$$

7 Design a push-pull PWM converter to meet the following specifications: the dc input voltage V_I is the US single-phase rectified utility line voltage, $V_O = 5\,\text{V}$, $I_O = 4$ to $40\,\text{A}$, and $V_r/V_O \leq 1\,\%$.

The maximum and minimum values of the dc output power are

$$P_{Omax} = V_O I_{Omax} = 5 \times 40 = 200\,\text{W}$$

and

$$P_{Omin} = V_O I_{Omin} = 5 \times 4 = 20\,\text{W}.$$

The minimum and maximum values of the load resistance are

$$R_{Lmin} = \frac{V_O}{I_{Omax}} = \frac{5}{40} = 0.125\,\Omega$$

and

$$R_{Lmax} = \frac{V_O}{I_{Omin}} = \frac{5}{4} = 1.25\,\Omega.$$

The minimum, nominal, and maximum values of the dc voltage transfer function are

$$M_{V\,DCmin} = \frac{V_O}{V_{Imax}} = \frac{5}{187} = 0.02674 = \frac{1}{37.4},$$

$$M_{V\,DCnom} = \frac{V_O}{V_{Inom}} = \frac{5}{156} = 0.03205 = \frac{1}{31.2},$$

and

$$M_{V\,DCmax} = \frac{V_O}{V_{Imin}} = \frac{5}{127} = 0.03937 = \frac{1}{25.4}.$$

Assume initially that $D_{max} \approx 0.4$ and $\eta = 80\,\%$. From (9.92), the transformer turn ratio is

$$n = \frac{2D_{max}\eta}{M_{V\,DCmax}} = \frac{2 \times 0.4 \times 0.8}{0.03937} = 16.256.$$

Let $n = 16$. The minimum, nominal, and maximum values of the duty cycle are

$$D_{min} = \frac{nM_{V\,DCmin}}{2\eta} = \frac{16 \times 0.02674}{2 \times 0.8} = 0.2674,$$

$$D_{nom} = \frac{nM_{V\,DCnom}}{2\eta} = \frac{16 \times 0.03205}{2 \times 0.8} = 0.3205,$$

and

$$D_{max} = \frac{nM_{V\,DCmax}}{2\eta} = \frac{16 \times 0.03937}{2 \times 0.8} = 0.3937.$$

Assume the switching frequency $f_s = 100\,\text{kHz}$. The minimum inductance required maintain the converter in CCM is

$$L_{min} = \frac{R_{Lmax}\left(\frac{1}{2} - D_{min}\right)}{2f_s} = \frac{1.25 \times \left(\frac{1}{2} - 0.2674\right)}{2 \times 100 \times 10^3} = 1.45375\,\mu\text{H}.$$

Pick $L = 4.7\,\mu\text{H}$.

The maximum peak-to-peak ripple of the inductor current is

$$\Delta i_{Lmax} = \frac{V_O\left(\frac{1}{2} - D_{min}\right)}{f_s L} = \frac{5 \times \left(\frac{1}{2} - 0.2674\right)}{100 \times 10^3 \times 4.7 \times 10^{-6}} = 2.47\,\text{A}.$$

The ripple voltage is

$$V_r = \frac{V_O}{100} = \frac{5}{100} = 50\,\text{mV}.$$

If the filter capacitance is large enough, $V_r = r_{Cmax}\,\Delta i_{Lmax}$ and the maximum ESR of filter capacitor is

$$r_{Cmax} = \frac{V_r}{\Delta i_{Lmax}} = \frac{50 \times 10^{-3}}{2.47} = 20.24\,\text{m}\Omega.$$

Pick a capacitor with $r_C = 20\,\text{m}\Omega$.

The minimum value of the filter capacitance at which the ripple voltage is determined by the ripple voltage across the ESR is

$$C_{min} = \max\left\{ \frac{D_{max}}{2f_s r_C}, \frac{\frac{1}{2} - D_{min}}{2f_s r_C} \right\} = \max\left\{ \frac{0.3937}{2f_s r_C}, \frac{\frac{1}{2} - 0.2674}{2f_s r_C} \right\}$$

$$= \frac{D_{max}}{2f_s r_C} = \frac{0.3937}{2 \times 100 \times 10^3 \times 20 \times 10^{-3}} = 98.425\,\mu\text{F}.$$

Pick $C = 100\,\mu\text{F}/20\,\text{m}\Omega/10\,\text{V}$.

The voltage and current stresses of the rectifier diodes are

$$V_{DMmax} = \frac{2V_{Imax}}{n} = \frac{2 \times 187}{16} = 23.375\,\text{V}$$

and

$$I_{DMmax} = I_{Omax} + \frac{\Delta i_{Lmax}}{2} = 40 + \frac{2.47}{2} = 41.235\,\text{A}.$$

The maximum peak current through the ideal transformer primary winding is

$$I_{1max} = \frac{I_{DMmax}}{n} = \frac{41.235}{16} = 2.577\,\text{A}.$$

Assuming that the peak-to-peak current through the upper magnetizing inductance is less than 10% of the maximum peak current through the ideal transformer primary winding, we have

$$\Delta i_{Lm1(max)} = 0.1 I_{1max} = 0.1 \times 2.577 = 0.258\,\text{A}.$$

Thus, the minimum magnetizing inductance is

$$L_{m1(min)} = \frac{D_{min} V_{Imax}}{f_s \Delta i_{Lm1(max)}} = \frac{0.2674 \times 187}{100 \times 10^3 \times 0.258} = 1.94\,\text{mH}.$$

Pick $L_{m1} = L_{m2} = 2\,\text{mH}$.

The voltage and current stresses of the power MOSFETs are

$$V_{SMmax} = 2V_{Imax} = 2 \times 187 = 374\,\text{V}$$

and

$$I_{SMmax} = \frac{I_{Omax}}{n} + \frac{\Delta i_{Lmax}}{2n} + \frac{\Delta i_{Lm1(max)}}{2}$$

$$= \frac{40}{16} + \frac{2.47}{2 \times 16} + \frac{0.26}{2} = 2.707\,\text{A}.$$

Motorola MTM4N45 power MOSFETs are selected, which have $V_{DSS} = 450\,\text{V}$, $I_{SM} = 4\,\text{A}$ of continuous current, $r_{DS} = 1.5\,\Omega$, $Q_g = 32\,\text{nC}$, and $C_o = 100\,\text{pF}$. International Rectifier 55HQ030 Schottky diodes are chosen, which have $I_{D(AV)max} = 60\,\text{A}$, $I_{FSM} = 300\,\text{A}$, $V_{DMmax} = 30\,\text{V}$, $V_F = 0.5\,\text{V}$, and $R_F = 14\,\text{m}\Omega$.

The conduction power loss in each MOSFET is

$$P_{rDS1} = \frac{D_{max} r_{DS} I_{Omax}^2}{n^2} = \frac{0.3937 \times 1.5 \times 40^2}{16^2} = 3.69\,\text{W}.$$

The switching loss associated with turning each transistor on and off is

$$P_{sw} = 4 f_s C_o V_{Imax}^2 = 4 \times 100 \times 10^3 \times 100 \times 10^{-12} \times 187^2 = 1.399\,\text{W}.$$

The overall power loss in each transistor is

$$P_{MOS} = P_{rDS1} + \frac{P_{sw}}{2} = 3.69 + \frac{1.399}{2} = 4.39\,\text{W}.$$

Assume that the resistances of the primary windings are $r_{T1} = r_{T2} = 25\,\text{m}\Omega$ and the resistances of the secondary windings are $r_{T3} = r_{T4} = 10\,\text{m}\Omega$. Then the conduction loss in the transformer windings are

$$P_{rT1} = \frac{D_{max} r_{T1} I_{Omax}^2}{n^2} = \frac{0.3937 \times 0.025 \times 40^2}{16^2} = 0.062\,\text{W}$$

and

$$P_{rT3} = \frac{(2 D_{max} + 1) r_{T3} I_{Omax}^2}{4} = \frac{(2 \times 0.3937 + 1) \times 0.01 \times 40^2}{4}$$
$$= 7.15\,\text{W}.$$

The conduction loss in the diode due to R_F is

$$P_{RF1} = \frac{(2 D_{max} + 1) R_F I_{Omax}^2}{4} = \frac{(2 \times 0.3937 + 1) \times 0.014 \times 40^2}{4}$$
$$= 10.01\,\text{W},$$

the conduction loss in the diode due to V_F is

$$P_{VF1} = \frac{V_F I_{Omax}}{2} = \frac{0.5 \times 40}{2} = 10\,\text{W},$$

so that the total conduction loss in each diode is

$$P_{D1} = P_{RF1} + P_{VF1} = 10.01 + 10 = 20.01\,\text{W}.$$

Assuming that the ESR of the dc inductor is $r_L = 10\,\text{m}\Omega$,

$$P_{rL} = r_L I_{Omax}^2 = 0.01 \times 40^2 = 16\,\text{W}$$

and the power loss in the capacitor ESR is

$$P_{rC} = \frac{r_C (\Delta i_{Lmax})^2}{12} = \frac{0.02 \times 2.47^2}{12} = 0.01\,\text{W}.$$

The total power loss is

$$P_{LS} = 2 P_{rDS} + 2 P_{sw} + 2 P_{rT1} + 2 P_{rT3} + 2 P_{D1} + P_{rL} + P_{rC}$$
$$= 2 \times 3.69 + 2 \times 1.399 + 2 \times 0.062 + 2 \times 7.15 + 2 \times 20.01 + 16 + 0.01$$
$$= 80.632\,\text{W},$$

and the efficiency of the converter at full load is

$$\eta = \frac{P_{Omax}}{P_{Omax} + P_{LS}} = \frac{200}{200 + 80.632} = 71.27\%.$$

The peak-to-peak gate-source voltage is $V_{GSpp} = 14\,V$. Hence, the gate-drive power per transistor is found to be

$$P_G = f_s V_{GSpp} Q_g = 100 \times 10^3 \times 14 \times 32 \times 10^{-9} = 44.8\,mW.$$

8   Design a push-pull converter to meet the following specifications: $V_I = 12\,V \pm 30\%$, $V_O = 48\,V$, $I_O = 5$ to $50\,A$, and $V_r/V_O \le 1\%$.

The maximum input voltage is

$$V_{Imax} = 12 + 0.3 \times 12 = 15.6\,V$$

and the minimum input voltage is

$$V_{Imin} = 12 - 0.3 \times 12 = 8.4\,V.$$

The maximum dc voltage transfer function is

$$M_{V\,DCmax} = \frac{V_O}{V_{Imin}} = \frac{48}{8.4} = 5.714$$

and the minimum dc voltage transfer function is

$$M_{V\,DCmin} = \frac{V_O}{V_{Imax}} = \frac{48}{15.6} = 3.077.$$

The maximum load resistance is

$$R_{Lmax} = \frac{V_O}{I_{Omin}} = \frac{48}{5} = 9.6\,\Omega.$$

Let $D_{max} = 0.4$ and $\eta = 0.9$.

$$n = \frac{2\eta D_{max}}{M_{V\,DCmax}} = \frac{2 \times 0.9 \times 0.4}{5.714} = 0.126 = \frac{1}{7.936}.$$

Pick $n = 1/8$. In this case,

$$D_{max} = \frac{n M_{V\,DCmax}}{2\eta} = \frac{(1/8) \times 5.714}{2 \times 0.9} = 0.3968$$

and

$$D_{min} = \frac{n M_{V\,DCmin}}{2\eta} = \frac{(1/8) \times 3.076}{2 \times 0.9} = 0.2136.$$

Hence,

$$L_{min} = \frac{R_{Lmax}(0.5 - D_{min})}{2 f_s} = \frac{9.6 \times (0.5 - 0.2136)}{2 \times 200 \times 10^3} = 6.8736\,\mu H.$$

Pick $L = 7\,\mu H$. Next,

$$\Delta i_{Lmax} = \frac{V_O(0.5 - D_{min})}{f_s L} = \frac{48(0.5 - 0.2136)}{200 \times 10^3 \times 7 \times 10^{-6}} = 9.82\,A.$$

The ripple voltage is

$$V_r = 0.01V_O = 0.01 \times 48 = 0.48 \text{ V}.$$

The maximum ESR resistance is

$$r_{Cmax} = \frac{V_r}{\Delta i_{Lmax}} = \frac{0.48}{9.82} = 48.87 \text{ m}\Omega.$$

Pick $r_C = 40 \text{ m}\Omega$. The minimum capacitance is

$$C_{min} = \max\left\{\frac{D_{max}}{2f_s r_C}, \frac{\frac{1}{2} - D_{min}}{2f_s r_C}\right\} = \max\left\{\frac{0.3968}{2f_s r_C}, \frac{\frac{1}{2} - 0.2136}{2f_s r_C}\right\}$$

$$= \frac{D_{max}}{2f_s r_C} = \frac{0.3968}{2 \times 200 \times 10^3 \times 40 \times 10^{-3}} = 24.8 \text{ }\mu\text{F}.$$

Pick $C = 50 \text{ }\mu\text{F}/100 \text{ V}/40 \text{ m}\Omega$.

Let us use a full-bridge rectifier. The voltage and current diode stresses are

$$V_{DMmax} = \frac{V_{Imax}}{n} = \frac{15.6}{1/8} = 124.8 \text{ V}$$

and

$$I_{DMmax} = I_{Omax} + \frac{\Delta i_{Lmax}}{2} = 50 + \frac{9.82}{2} = 54.91 \text{ A}.$$

The voltage and current switch stresses are

$$V_{SMmax} = 2V_{Imax} = 2 \times 15.6 = 31.2 \text{ V},$$

$$I_{1max} = \frac{I_{DMmax}}{n} = \frac{54.91}{1/8} = 439.28 \text{ A},$$

$$\Delta i_{Lm1(max)} = 0.1 I_{1max} = 43.92 \text{ A},$$

$$I_{SMmax} = \frac{I_{DMmax}}{n} + \frac{\Delta i_{Lmax}}{2n} + \frac{\Delta i_{Lm1(max)}}{2}$$

$$= \frac{54.91}{1/8} + \frac{9.82}{2 \times (1/8)} + \frac{43.92}{2} = 500.52 \text{ A}.$$

9.9 In a push-pull converter V_I is the US single-phase rectified utility line, $V_O = 12$ $I_O = 0.2$ to 2 A, and $f_s = 200 \text{ kHz}$. Find the maximum inductance for DCM operation

The maximum input voltage is

$$V_{Imax} = 90\sqrt{2} = 127 \text{ V}.$$

The minimum load resistance is

$$R_{Lmin} = \frac{V_O}{I_{Omax}} = \frac{12}{2} = 6 \text{ }\Omega.$$

The maximum dc voltage transfer function is

$$M_{V\,DCmax} = \frac{V_O}{V_{Imax}} = \frac{12}{127} = 0.0945.$$

Let $D_{max} = 0.4$ and $\eta = 0.9$. Hence,

$$n = \frac{2\eta D_{max}}{M_{V\,DCmax}} = \frac{2 \times 0.9 \times 0.4}{0.0945} = 7.619.$$

Pick $n = 7$. Thus,

$$D_{max} = \frac{nM_{V\,DCmax}}{2\eta} = \frac{7 \times 0.0945}{2 \times 0.9} = 0.3663.$$

Finally,

$$L_{max} = \frac{R_{Lmin}(0.5 - D_{max})}{2f_s} = \frac{6 \times (0.5 - 0.3663)}{2 \times 200 \times 10^3} = 2.0055\,\mu H.$$

Pick $L = 1.5\,\mu H.$

Chapter 10

Small-signal Models of PWM Converters for CCM and DCM

10.1 A single-ended transformerless PWM converter is operated under steady-state conditions in CCM. The duty cycle is $D = 0.4$. Find the components of an averaged model of the ideal switching network.

The averaged dc model of the ideal switching network consists of a dependent current source (Figure 10.4(c), right)

$$I_S = DI_L = 0.4I_L$$

and a dependent voltage source (Figure 10.4(c), right)

$$V_{LD} = DV_{SD} = 0.4V_{SD}.$$

10.2 The parasitic components of a single-ended transformerless PWM converter operated under steady-state conditions in CCM are $r_{DS} = 1\,\Omega$, $R_F = 24\,\text{m}\Omega$, $V_F = 0.7\,\text{V}$, and the duty cycle is $D = 0.4$. Find the values of the averaged parasitic components in the original branches.

The averaged on-resistance of the MOSFET placed in the switch branch is

$$r_{DS\,AV(S)} = \frac{r_{DS}}{D} = \frac{1}{0.4} = 2.5\,\Omega.$$

The averaged forward resistance of the diode placed in the diode branch is

$$R_{F\,AV(D)} = \frac{R_F}{1 - D} = \frac{0.024}{1 - 0.4} = 40\,\text{m}\Omega.$$

The averaged offset voltage of the diode placed in the diode branch is

$$V_{F\,AV(D)} = V_F = 0.7\,\text{V}.$$

10.3 The parasitic components of a single-ended transformerless PWM converter operate under steady-state conditions in CCM are $r_{DS} = 1\,\Omega$, $R_F = 24\,\text{m}\Omega$, $V_F = 0.7\,\text{V}$, an the duty cycle is $D = 0.4$. Find the values of the averaged parasitic components in t inductor branch.

The averaged on-resistance of the MOSFET placed in the inductor branch is

$$r_{DS\,AV(L)} = D r_{DS} = 0.4 \times 1 = 0.4\,\Omega.$$

The averaged forward resistance of the diode placed in the inductor branch is

$$R_{F\,AV(L)} = (1 - D)R_F = (1 - 0.4) \times 0.024 = 14.4\,\text{m}\Omega.$$

The averaged offset voltage of the diode placed in the inductor branch is

$$V_{F\,AV(L)} = (1 - D)V_F = (1 - 0.4) \times 0.7 = 0.42\,\text{V}.$$

10.4 The parasitic components of a single-ended transformerless PWM converter operat under steady-state conditions in CCM are $r_{DS} = 1\,\Omega$, $R_F = 24\,\text{m}\Omega$, $V_F = 0.7\,\text{V}$, a the duty cycle is $D = 0.4$. Find the values of the averaged parasitic components in t switch branch.

The averaged on-resistance of the MOSFET placed in the switch branch is

$$r_{DS\,AV(S)} = \frac{r_{DS}}{D} = \frac{1}{0.4} = 2.5\,\Omega.$$

The averaged forward resistance of the diode placed in the switch branch is

$$R_{F\,AV(S)} = \frac{(1 - D)R_F}{D^2} = \frac{(1 - 0.4) \times 0.024}{0.4^2} = 90\,\text{m}\Omega.$$

The averaged offset voltage of the diode placed in the switch branch is

$$V_{F\,AV(S)} = \frac{(1 - D)V_F}{D} = \frac{(1 - 0.4) \times 0.7}{0.4} = 1.05\,\text{V}.$$

10.5 The parasitic components of a single-ended transformerless PWM converter operat under steady-state conditions in CCM are $r_{DS} = 1\,\Omega$, $R_F = 24\,\text{m}\Omega$, $V_F = 0.7\,\text{V}$, an the duty cycle is $D = 0.4$. Find the values of the averaged parasitic components in diode branch.

The averaged on-resistance of the MOSFET placed in the diode branch is

$$r_{DS\,AV(D)} = \frac{D r_{DS}}{(1 - D)^2} = \frac{0.4 \times 1}{(1 - 0.4)^2} = 1.1111\,\Omega.$$

The averaged forward resistance of the diode placed in the diode branch is

$$R_{F\,AV(D)} = \frac{R_F}{1 - D} = \frac{0.024}{1 - 0.4} = 40\,\text{m}\Omega.$$

The averaged offset voltage of the diode placed in the diode branch is

$$V_{F\,AV(D)} = V_F = 0.7\,\text{V}.$$

.6 The parasitic components of a single-ended transformerless PWM converter operated under steady-state conditions in CCM are $r_{DS} = 1\,\Omega$, $R_F = 24\,\text{m}\Omega$, $V_F = 0.7\,\text{V}$, $r_L = 0.2\,\Omega$, and the duty cycle is $D = 0.4$. Find the total averaged parasitic resistance in the inductor branch.

The total averaged parasitic resistance reflected to the inductor branch is

$$r = Dr_{DS} + (1 - D)R_F + r_L$$

$$= 0.4 \times 1 + (1 - 0.4) \times 0.024 + 0.2 = 0.4 + 0.0144 + 0.2 = 0.6144\,\Omega.$$

.7 The duty cycle of a single-ended transformerless PWM converter operated under steady-state conditions in CCM is $D = 0.4$ and the ESR of the inductor is $r_L = 0.2\,\Omega$. Find the resistance of the inductor ESR in (a) the switch branch and (b) the diode branch.

The inductor resistance shifted to the switch branch is

$$r_{L(S)} = \frac{r_L}{D^2} = \frac{0.2}{0.4^2} = 1.25\,\Omega.$$

The inductor resistance shifted to the diode branch is

$$r_{L(D)} = \frac{r_L}{(1 - D)^2} = \frac{0.2}{(1 - 0.4)^2} = 0.556\,\Omega.$$

.8 A single-ended transformerless PWM converter operated under steady-state conditions in CCM has $D = 0.4$, $V_{SD} = 24\,\text{V}$, $I_L = 1.8\,\text{A}$, $r_{DS} = 1\,\Omega$, $R_F = 24\,\text{m}\Omega$, $V_F = 0.7\,\text{V}$, and $r_L = 0.2\,\Omega$. Find the components of the small-signal model of the actual switching network with all parasitic elements in the inductor branch.

The small-signal model consists of two dependent current sources $I_L d$ and Di_l, two dependent voltage sources $V_{SD}d$ and Dv_{sd} and a resistance r. The current-controlled current source is

$$Di_l = 0.4i_l\,\text{A}.$$

The duty-cycle-controlled current source is

$$I_L d = 1.8d\,\text{A}.$$

The voltage-controlled voltage source is

$$Dv_{sd} = 0.4v_{sd}\,\text{V}.$$

The duty-cycle-controlled voltage source is

$$V_{SD}d = 24d\,\text{V}.$$

The total averaged parasitic resistance reflected to the inductor branch is

$$r = Dr_{DS} + (1 - D)R_F + r_L$$

$$= 0.4 \times 1 + (1 - 0.4) \times 0.024 + 0.2 = 0.4 + 0.0144 + 0.2 = 0.6144\,\Omega.$$

Chapter 11

Open-loop Small-signal Characteristics of Boost Converter for CCM

.1   The boost converter designed in Chapter 3 has $V_{Inom} = 156\,\text{V}$, $V_O = 400\,\text{V}$, $D_{nom} = 0.65$, $R_{Lmin} = 1.778\,\text{k}\Omega$, $r_{DS} = 1\,\Omega$, $V_F = 1.4\,\text{V}$, $R_F = 0.0171\,\Omega$, $L = 30\,\text{mH}$, $r_L = 2.1\,\Omega$, $C = 1\mu\text{F}$, and $r_C = 1\,\Omega$. Determine $M_{V\,DC}$ and η.

The total averaged parasitic resistance placed in the inductor branch is

$$r = D_{nom}r_{DS} + (1 - D_{nom})R_F + r_L$$

$$= 0.65 \times 1 + (1 - 0.65) \times 0.0171 + 2.1 = 0.65 + 0.006 + 2.1$$

$$= 2.756\,\Omega.$$

The dc input-to-output transfer function is

$$M_{V\,DC} = \frac{1}{1 - D_{nom}} \frac{1}{1 + \dfrac{V_F}{V_O} + \dfrac{r}{(1 - D_{nom})^2 R_{Lmin}}}$$

$$= \frac{1}{1 - 0.65} \frac{1}{1 + \dfrac{1.4}{400} + \dfrac{2.756}{(1 - 0.65)^2 \times 1778}}$$

$$= 2.8395 = 9.06\,\text{dB}.$$

The efficiency of the converter is

$$\eta = \frac{1}{1 + \dfrac{V_F}{V_O} + \dfrac{r}{(1 - D_{nom})^2 R_{Lmin}}}$$

$$= \frac{1}{1 + \dfrac{1.4}{400} + \dfrac{2.756}{(1 - 0.65)^2 \times 1778}} = 98.38\,\%.$$

11.2 The boost converter has $V_{Omin} = 156\,\text{V}$, $V_O = 400\,\text{V}$, $D_{nom} = 0.65$, $R_{Lmin} = 1.778\,\text{k}\Omega$, $r_{DS} = 1\Omega$, $R_F = 0.0171\Omega$, $L = 30\,\text{mH}$, $r_L = 2.1\Omega$, $C = 1\mu\text{F}$, $r = 2.756\,\Omega$, and $r_C = 1\,\Omega$. Calculate z_n, f_{zn}, z_p, f_{zp}, f_0, ξ, Q, p_1, p_2, and f_d.

The LHP zero is

$$z_n = -\frac{1}{r_C C} = -\frac{1}{1 \times 1 \times 10^{-6}} = -10^6 \,\text{rad/s},$$

and the frequency of the LHP zero is

$$f_{zn} = -\frac{z_n}{2\pi} = -\frac{-10^6}{2\pi} = 159\,\text{kHz}.$$

The RHP zero is

$$z_p = \omega_{zp} = \frac{1}{L}[R_{Lmin}(1 - D_{nom})^2 - r]$$

$$= \frac{1}{30 \times 10^{-3}}[1778 \times (1 - 0.65)^2 - 2.756] = 7168.3\,\text{rad/s},$$

and the frequency of the RHP zero is

$$f_{zp} = \frac{\omega_{zp}}{2\pi} = \frac{7168.3}{2\pi} = 1.14\,\text{kHz}.$$

The corner frequency is

$$f_0 = \frac{1}{2\pi}\sqrt{\frac{R_{Lmin}(1 - D_{nom})^2 + r}{LC(R_{Lmin} + r_C)}}$$

$$= \frac{1}{2\pi}\sqrt{\frac{1778 \times (1 - 0.65)^2 + 2.756}{30 \times 10^{-3} \times 1 \times 10^{-6} \times (1778 + 1)}} = 323.55\,\text{Hz}.$$

The damping ratio is

$$\xi = \frac{C[r(R_{Lmin} + r_C) + (1 - D_{nom})^2 R_{Lmin}r_C] + L}{2\sqrt{LC(R_{Lmin} + r_C)}[r + (1 - D_{nom})^2 R_{Lmin}]}$$

$$= \frac{1 \times 10^{-6}[2.756 \times (1778 + 1) + (1 - 0.65)^2 \times 1778 \times 1] + 30 \times 10^{-3}}{2\sqrt{30 \times 10^{-3} \times 1 \times 10^{-6} \times (1778 + 1)}[2.756 + (1 - 0.65)^2 \times 1778]}$$

$$= \frac{35.126 \times 10^{-3}}{216.99 \times 10^{-3}} = 0.162,$$

and the quality factor is

$$Q = \frac{1}{2\xi} = \frac{1}{2 \times 0.162} = 3.086.$$

The poles are

$$p_1, p_2 = -\sigma \pm j\omega_d = -\xi\omega_0 \pm j\omega_0\sqrt{1-\xi^2}$$

$$= -0.162 \times 2 \times \pi \times 323.55 \pm j2 \times \pi \times 323.55 \times \sqrt{1 - 0.162^2}$$

$$= -329.3 \pm j2006.07 \, \text{rad/s},$$

and the damped frequency is

$$f_d = \frac{\omega_d}{2\pi} = \frac{2006.07}{2\pi} = 319.27 \, \text{Hz}.$$

1.3   The boost converter has $V_{Inom} = 156\,\text{V}$, $V_O = 400\,\text{V}$, $D_{nom} = 0.65$, $R_{Lmin} = 1.778\,\text{k}\Omega$, $r_{DS} = 1\,\Omega$, $R_F = 0.0171\,\Omega$, $L = 30\,\text{mH}$, $r_L = 2.1\,\Omega$, $C = 1\,\mu\text{F}$, $r = 2.756\,\Omega$, and $r_C = 1\,\Omega$. Determine T_{po} and $T_p(\infty)$.

The control-to-output transfer function at $f = 0$ is

$$T_{po} = \frac{V_O}{1 - D_{nom}} \frac{R_{Lmin}(1 - D_{nom})^2 - r}{R_{Lmin}(1 - D_{nom})^2 + r}$$

$$= \frac{400}{1 - 0.65} \frac{1778 \times (1 - 0.65)^2 - 2.756}{1778 \times (1 - 0.65)^2 + 2.756}$$

$$= 1128.5\,\text{V} = 61.05\,\text{dBV},$$

and

$$T_p(\infty) = -\frac{V_O}{1 - D_{nom}} \frac{r_C}{R_{Lmin} + r_C} = -\frac{400}{1 - 0.65} \frac{1}{1778 + 1} = -0.6424\,\text{V},$$

which gives

$$|T_p(\infty)| = -3.84\,\text{dBV}.$$

1.4   The boost converter has $V_{Inom} = 156\,\text{V}$, $V_O = 400\,\text{V}$, $D_{nom} = 0.65$, $R_{Lmin} = 1.778\,\text{k}\Omega$, $r_{DS} = 1\,\Omega$, $R_F = 0.0171\,\Omega$, $L = 30\,\text{mH}$, $r_L = 2.1\,\Omega$, $C = 1\,\mu\text{F}$, $r = 2.756\,\Omega$, and $r_C = 1\,\Omega$. Determine M_{vo}.

The input-to-output transfer function at $f = 0$ is

$$M_{vo} = \frac{1}{1 - D_{nom}} \frac{R_{Lmin}}{R_{Lmin} + \dfrac{r}{(1 - D_{nom})^2}}$$

$$= \frac{1}{1 - 0.65} \frac{1778}{1778 + \dfrac{2.756}{(1 - 0.65)^2}} = 2.8214 = 9\,\text{dB}.$$

11.5 The boost converter has $V_{Inom} = 150\,V$, $V_O = 400\,V$, $D_{nom} = 0.65$, $R_{Lmin} = 1.778\,k\Omega$ $r_{DS} = 1\,\Omega$, $R_F = 0.0171\,\Omega$, $L = 30\,\mu H$, $r_L = 2.1\,\Omega$, $C = 1\,\mu F$, $r = 2.756\,\Omega$, and r_C $1\,\Omega$. Determine $Z_i(0)$.

The open-loop input impedance at $f = 0$ is

$$Z_i(0) = R_{Lmin}(1 - D_{nom})^2 + r = 1778 \times (1 - 0.65)^2 + 2.579 = 220.384\,\Omega.$$

11.6 The boost converter has $V_{Inom} = 28\,V$, $V_O = -12\,V$, $D_{nom} = 0.65$, $R_{Lmin} = 1.778\,k\Omega$ $r_{DS} = 1\,\Omega$, $R_F = 0.0171\,\Omega$, $L = 30\,mH$, $r_L = 2.1\,\Omega$, $C = 1\,\mu F$, $r = 2.756\,\Omega$, and r_C $1\,\Omega$. Determine $Z_o(0)$ and $Z_o(\infty)$.

The open-loop output impedance at $f = 0$ is

$$Z_o(0) = \frac{rR_{Lmin}}{R_{Lmin}(1 - D_{nom})^2 + r} = \frac{2.756 \times 1778}{1778 \times (1 - 0.65)^2 + 2.756}$$

$$= 22.216\,\Omega,$$

and we have

$$Z_o(\infty) = \frac{R_{Lmin}r_C}{R_{Lmin} + r_C} = \frac{1778 \times 1}{1778 + 1} \approx 1\,\Omega.$$

11.7 The boost converter has $R_{Lmin} = 1.778\,\Omega$, $r_{DS} = 1\,\Omega$, $R_F = 0.0171\,\Omega$, $L = 30\,mH$, r_L $2.1\,\Omega$, $C = 1\,\mu F$, $r = 2.756\,\Omega$, and $r_C = 1\,\Omega$. Determine $Z_o(0)$ for $D = 0.1, 0.5, 0$ and 0.9.

Assume that r is independent of D. For $D = 0.1$,

$$Z_o(0) = \frac{rR_{Lmin}}{R_{Lmin}(1 - D)^2 + r} = \frac{2.756 \times 1778}{1778 \times (1 - 0.1)^2 + 2.756} = 3.4\,\Omega.$$

For $D = 0.5$,

$$Z_o(0) = \frac{rR_{Lmin}}{R_{Lmin}(1 - D)^2 + r} = \frac{2.756 \times 1778}{1778 \times (1 - 0.5)^2 + 2.756} = 10.96\,\Omega.$$

For $D = 0.8$,

$$Z_o(0) = \frac{rR_{Lmin}}{R_{Lmin}(1 - D)^2 + r} = \frac{2.756 \times 1778}{1778 \times (1 - 0.8)^2 + 2.756} = 66.329\,\Omega.$$

For $D = 0.9$,

$$Z_o(0) = \frac{rR_{Lmin}}{R_{Lmin}(1 - D)^2 + r} = \frac{2.756 \times 1778}{1778 \times (1 - 0.9)^2 + 2.756} = 238.61\,\Omega.$$

11.8 The boost converter has $V_{Inom} = 156\,V$, $V_O = 400\,V$, $D_{nom} = 0.65$, $R_{Lmin} = 1.778\,k$ $r_{DS} = 1\,\Omega$, $R_F = 0.0171\,\Omega$, $L = 30\,mH$, $r_L = 2.1\,\Omega$, $C = 1\,\mu F$, $r = 2.756\,\Omega$, and r_C $1\,\Omega$. Determine ξ and Q at all parasitic resistances equal to zero. Calculate the ratio the values of ξ.

The approximate value of the damping ratio is

$$\xi_{lossless} \approx \frac{1}{2R_{Lmin}}\sqrt{\frac{L}{C(1-D_{nom})^2}} = \frac{1}{2 \times 1778}\sqrt{\frac{30 \times 10^{-3}}{1 \times 10^{-6} \times (1-0.65)^2}}$$
$$= 0.139,$$

resulting in the approximate value of the loaded quality factor

$$Q = \frac{1}{2\xi} = \frac{1}{2 \times 0.139} = 3.597.$$

Hence, the ratio of the actual and approximate values of ξ is

$$\frac{\xi_{lossy}}{\xi_{lossless}} = \frac{0.162}{0.139} = 1.165.$$

In this case, the two values are close to each other.

Chapter 12

Voltage-mode Control of Boost Converter

.1 A pulse-width modulator has a ramp output voltage with a peak value $V_{Tm} = 5\,\text{V}$. Find a transfer function of the modulator.

The transfer function of the PWM modulator is

$$T_m = \frac{1}{V_{Tm}} = \frac{1}{5} = 0.2\,\text{V}^{-1}.$$

.2 A boost PWM converter has $V_{Inom} = 156\,\text{V}$, $D_{nom} = 0.65$, and the pulse-width modulator $V_{Tm} = 5\,\text{V}$. Determine the dc reference voltage for a control circuit.

The reference voltage is

$$V_R = D_{nom} V_{Tm} = 0.65 \times 5 = 3.25\,\text{V}.$$

.3 A boost PWM converter has $V_O = 400\,\text{V}$ and the reference voltage is $V_R = 3.25\,\text{V}$. Design a feedback network.

The voltage transfer function of the feedback network is

$$\beta = \frac{V_R}{V_O} = \frac{3.25}{400} = 0.008125 = \frac{1}{123} = -41.8\,\text{dB}.$$

The transfer function of the feedback network is given by

$$\beta = \frac{V_R}{V_O} = \frac{R_B}{R_A + R_B}.$$

Let $R_B = 1\,\text{k}\Omega$. Hence,

$$R_A = R_B \left(\frac{1}{\beta} - 1 \right) = 1 \times \left(\frac{1}{1/123} - 1 \right) = 122\,\text{k}\Omega.$$

Pick $R_A = 120 \text{k}\Omega$. Thus,

$$h_{11} = \frac{R_A R_B}{R_A + R_B} = \frac{1 \times 120}{1 + 120} = 992\,\Omega.$$

12.4   A boost PWM converter has $T_{po} = 1114.27\,\text{V}$, a feedback network has $\beta = 1/12$ and a pulse-width modulator has $T_m = 0.2\,\text{V}^{-1}$. Determine the overall voltage transf function T_{ko} of the three stages at $f = 0$.

At $f = 0$,

$$T_{ko} = \beta T_m T_{po} = \frac{1}{123} \times 0.2 \times 1114.27 = 1.8118 = 5.162\,\text{dB}.$$

12.5   A boost PWM converter has $V_I = 156\,\text{V}$, $V_O = 400\,\text{V}$, $V_R = 3.25\,\text{V}$, $T_{ko} = 1.81$ $\xi = 0.162$, $f_{zn} = 159\,\text{kHz}$, $f_{zp} = 1.17\,\text{kHz}$, and $f_0 = 322\,\text{Hz}$. Design a control circ such that $PM \geq 55°$.

Assume that $f_c = f_m = 0.5\,\text{kHz}$. From (12.19),

$$\phi_{\beta T1}(f_c) = -180° + \arctan\left(\frac{f_c}{f_{zn}}\right) - \arctan\left(\frac{f_c}{f_{zp}}\right) - \arctan\left[\frac{\left(\dfrac{2\xi f_c}{f_0}\right)}{1 - \left(\dfrac{f_c}{f_0}\right)^2}\right]$$

$$= -180° + \arctan\left(\frac{0.5}{159}\right) - \arctan\left(\frac{0.5}{1.17}\right)$$

$$- \arctan\left[\frac{\left(\dfrac{2 \times 0.162 \times 0.5}{0.322}\right)}{1 - \left(\dfrac{0.5}{0.322}\right)^2}\right]$$

$$= -180° + 0.18° - 23.14° + 19.62° = -183.34°.$$

Since the phase margin $PM = 60°$, one obtains the required phase boost

$$\phi_m = PM - \phi_{\beta T1}(f_c) - 90° = 60° + 183.34° - 90° = 153.34°,$$

and the K factor

$$K = \tan^2\left(\frac{\phi_m}{4} + 45°\right) = \tan^2\left(\frac{153.34°}{4} + 45°\right) = 73.23.$$

Next

$$|T_{ko}(f_c)| = T_{ko} \frac{\sqrt{1 + \left(\dfrac{\omega_c}{\omega_{zn}}\right)^2}\sqrt{1 + \left(\dfrac{\omega_c}{\omega_{zp}}\right)^2}}{\sqrt{\left[1 - \left(\dfrac{\omega_c}{\omega_0}\right)^2\right]^2 + \left(\dfrac{2\xi\omega_c}{\omega_0}\right)^2}}$$

$$= 1.8118 \frac{\sqrt{1 + \left(\dfrac{0.5}{159}\right)^2}\sqrt{1 + \left(\dfrac{0.5}{1.17}\right)^2}}{\sqrt{\left[1 - \left(\dfrac{0.5}{0.322}\right)^2\right]^2 + \left(\dfrac{2 \times 0.162 \times 0.5}{0.322}\right)^2}}$$

$$= 1.31558 = 2.823\,\text{dB}.$$

Assuming $R_1 = 100\,\text{k}\Omega$,

$$R_3 = \frac{R_1[R_1 - h_{11}(K - 1)]}{(K - 1)(R_1 + h_{11})} = \frac{100[100 - (73.23 - 1)0.992]}{(73.23 - 1)(100 + 0.992)}$$

$$= 0.3886\,\text{k}\Omega.$$

Pick $R_3 = 390\,\Omega$. Now,

$$C_2 = \frac{|T_{ko}(f_c)|}{\omega_c(R_1 + h_{11})} = \frac{1.31558}{2 \times \pi \times 0.5 \times 10^3 \times (100 + 0.992) \times 10^3}$$

$$= 4.14\,\text{nF}.$$

Pick $C_2 = 3.9\,\text{nF}$. Next,

$$C_1 = C_2(K - 1) = 3.9 \times (73.23 - 1) = 281.697\,\text{nF}.$$

Pick $C_1 = 270\,\text{nF}$. Then

$$R_2 = \frac{\sqrt{K}}{\omega_c C_1} = \frac{\sqrt{73.23}}{2 \times \pi \times 0.5 \times 10^3 \times 270 \times 10^{-9}} = 10.08\,\text{k}\Omega.$$

Pick $R_2 = 10\,\text{k}\Omega$. Finally,

$$C_3 = \frac{R_1 + h_{11}}{\omega_c\sqrt{K}[R_1R_3 + h_{11}(R_1 + R_3)]}$$

$$= \frac{(100 + 0.992) \times 10^3}{2 \times \pi \times 0.5 \times 10^3 \sqrt{73.23}[100 \times 0.27 + 0.992(100 + 0.27)] \times 10^6}$$

$$= 29.9\,\text{nF}.$$

Pick $C_3 = 33\,\text{nF}$.

12.6 The transfer function of the feedback network in a boost converter is $\beta = 1/123$. Find the value of the closed-loop control-to-input voltage transfer function at $f = 0$.

The closed-loop control-to-output transfer voltage function at $f = 0$ is

$$T_{clo} \approx \frac{1}{\beta} = \frac{1}{1/123} = 123 = 41.8\,\text{dB}.$$

12.7 A closed-loop boost converter has $D_{nom} = 0.65$, $R_{Lmin} = 1.778\,\text{k}\Omega$, $R_{Lmax} = 17.78\,\text{k}\Omega$ and $r = 2.756\,\Omega$. Find the values of the dc closed-loop input resistance at R_{Lmin} and R_{Lmax}.

At R_{Lmin},

$$R_{icl}(0) = -[(1 - D_{nom})^2 R_{Lmin} - r] = -[(1 - 0.65)^2 \times 1778 - 2.756]$$
$$= -215\,\Omega.$$

At R_{Lmax},

$$R_{icl}(0) = -[(1 - D_{nom})^2 R_{Lmax} - r] = -[(1 - 0.65)^2 \times 17{,}780 - 2.756]$$
$$= -2.175\,\text{k}\Omega.$$

Chapter 13
Current-mode Control

13.1 A lossless buck converter with constant-frequency peak-current-mode control and without slope compensation has $V_O = 5\,\text{V}$. What is the range of the input voltage V_I in which the inner loop is stable?

The dc voltage transfer function for the lossless buck converter is

$$M_{V\,DC} = \frac{V_O}{V_I} = D < 0.5.$$

Hence,

$$V_I > 2V_O = 2 \times 5 = 10\,\text{V}.$$

13.2 A lossy buck converter with constant-frequency peak-current-mode control and without slope compensation has $V_O = 5\,\text{V}$ and efficiency $\eta = 0.9$. What is the range of the input voltage V_I in which the converter is stable?

The dc voltage transfer function for the lossy buck converter is

$$M_{V\,DC} = \frac{V_O}{V_I} = \eta D.$$

The maximum dc voltage transfer function for the lossless buck converter is

$$M_{V\,DCmax} = \frac{V_O}{V_{Imin}} = \eta D_{max} = 0.9 \times 0.5 = 0.45.$$

Hence,

$$V_I > \frac{V_O}{\eta D_{max}} = \frac{5}{0.45} = 11.11\,\text{V}.$$

13.3   A lossy buck converter with constant-frequency peak-current-mode control and wi slope compensation has $V_I = 28\,\text{V}$, $V_O = 5\,\text{V}$, $L = 301\,\mu\text{H}$, the switching frequenc $f_s = 100\,\text{kHz}$, and efficiency $\eta = 0.9$. Is the convester stable? What is the slope of th rising inductor current waveform?

The dc voltage transfer function for the lossy buck converter is

$$M_{V\,DC} = \frac{V_O}{V_I} = \eta D.$$

The duty cycle is

$$D = \frac{V_O}{\eta V_I} = \frac{5}{0.9 \times 28} = 0.1984.$$

Thus, the converter is stable because $D < 0.5$. The slope of the rising inductor curre waveform is

$$M_1 = \frac{V_I - V_O}{L} = \frac{28 - 5}{301 \times 10^{-6}} = 76.41 \times 10^3\,\text{A/s}.$$

13.4   A buck converter with constant-frequency peak-current-mode control has $V_I = 28$ $4\,\text{V}$, $V_O = 2\,\text{V}$, $L = 301\,\mu\text{H}$, $f_s = 100\,\text{kHz}$, and efficiency $\eta = 1$. Is slope compensati required in this converter?

The dc voltage transfer function for the lossy buck converter is

$$M_{V\,DCmax} = \frac{V_O}{V_{Imin}} = D_{max}.$$

The maximum duty cycle for the lossy buck converter is

$$D_{max} = \frac{V_O}{V_{Imin}} = \frac{20}{24} = 0.8333.$$

Thus, it is necessary to use slope compensation because $D_{max} > 0.5$. The minimu slope of the rising inductor current waveform is

$$M_{1min} = \frac{V_{Imin} - V_O}{L} = \frac{24 - 20}{301 \times 10^{-6}} = 13.289 \times 10^3\,\text{A/s}.$$

Hence, the minimum slope of the compensating ramp is

$$M_{3min} = M_{1min}\left(\frac{D_{max} - 0.5}{1 - D_{max}}\right) = 13.289 \times 10^3\left(\frac{0.8333 - 0.5}{1 - 0.8333}\right)$$

$$= 26.578 \times 10^3\,\text{A/s}.$$

To achieve a margin of stability,

$$M_3 = 1.5 M_{3min} = 1.5 \times 27.6497 \times 10^3 = 39.975 = 39.867 \times 10^3\,\text{A/s}.$$

Pick $M_3 = 40 \times 10^3\,\text{V/s}$. The peak value of the compensating ramp voltage is

$$V_{Tm} = M_3 T_s = \frac{M_3}{f_s} = \frac{40 \times 10^3}{10^5} = 0.4\,\text{V}.$$

3.5 A buck converter with constant-frequency peak-current-mode control has $V_{Inom} = 28$ V, $V_O = 20$ V, $L = 301$ μH, $f_s = 100$ kHz, and efficiency $\eta = 1$. Find the ramp slope for the optimum compensation and the peak compensating voltage.

The nominal duty cycle is

$$D_{nom} = \frac{V_O}{V_{Inom}} = \frac{20}{28} = 0.7143.$$

The nominal slope of the rising inductor current is

$$M_{1nom} = \frac{V_{Inom} - V_O}{L} = \frac{28 - 20}{301 \times 10^{-6}} = 26.578 \times 10^3 \text{ A/s}.$$

The nominal slope of the falling inductor current is

$$M_{2nom} = \frac{D_{nom}}{1 - D_{nom}} M_{1nom} = \frac{0.7143}{1 - 0.7143} \times 26.578 \times 10^3$$
$$= 66.45 \times 10^3 \text{ A/s}.$$

For the optimum slope compensation,

$$M_{3opt} = M_{2nom} = 66.45 \times 10^3 \text{ A/s}.$$

The peak value of the compensating ramp voltage is

$$V_{Tm} = M_{3opt} T_s = \frac{M_{3opt}}{f_s} = \frac{66.45 \times 10^3}{10^5} = 0.664 \text{ V}.$$

The nominal transfer function of the pulse-width modulator for the optimum slope compensation is

$$T_m = \frac{f_s}{M_{1nom} + M_{3opt}} = \frac{10^5}{(26.578 + 66.45) \times 10^3} = 1.075 \text{ V}^{-1}$$
$$= 0.628 \text{ dB V}^{-1}.$$

3.6 A buck-boost converter with constant-frequency peak-current-mode control has $V_{Inom} = 42$ V, $V_O = -28$ V, $L = 334$ μH, $f_s = 100$ kHz, and efficiency $\eta = 0.85$. Find M_{3nom} and the peak value of the compensation ramp slope at which $a = 0.3$.

The nominal duty cycle is

$$D_{nom} = \frac{1}{1 - \dfrac{\eta}{M_{V\,DCnom}}} = \frac{1}{1 - \dfrac{\eta V_{Inom}}{V_O}} = \frac{1}{1 - \dfrac{0.85 \times 42}{-28}} = 0.43956.$$

The nominal slope of the rising inductor current is

$$M_{1nom} = \frac{V_{Inom}}{L} = \frac{42}{334 \times 10^{-6}} = 125,748.503 \text{ A/s},$$

and the nominal slope of the falling inductor current is

$$M_{2nom} = \frac{D_{nom}}{1 - D_{nom}} M_{1nom} = \frac{0.43956}{1 - 0.43956} \times 125,748.503 = 98,626.1 \, \text{A/s}.$$

Hence,

$$M_{3nom} = \frac{M_{2nom} - aM_{1nom}}{1 + a} = \frac{98,626.1 - 0.3 \times 125,748.503}{1 + 0.3} = 46,847.346 \, \text{A/s}.$$

Assuming $R_s = 1\Omega$, the peak value of the compensation ramp current is

$$I_{pk} = M_3 T_s = \frac{M_3}{f_s} = \frac{46,847.346}{10^5} = 0.468 \text{A}.$$

13.7 A PWM converter with constant-frequency peak-current-mode control has $R_s = 0.1 \, \Omega$, $a = 0.3$, and $f_s = 100 \, \text{kHz}$. Find $H_{icl}(z)$, $H_{icl}(s)$, and $T_i(s)$.

The closed-loop control voltage-to-inductor current transfer function in the z-domain

$$H_{icl}(z) = \frac{1 + a}{R_s} \frac{z}{z + a} = \frac{1 + 0.3}{0.1} \frac{z}{z + 0.3} = \frac{13z}{z + 0.3}.$$

The closed-loop control voltage-to-inductor current transfer function in the s-domain

$$H_{icl}(s) \approx \frac{1}{R_s} \frac{12 f_s^2}{s^2 + \frac{1 - a}{1 + a} 6 f_s s + 12 f_s^2} = \left(\frac{1}{0.1}\right) \frac{12 \times 10^{10}}{s^2 + \frac{1 - 0.3}{1 + 0.3} \times 6 \times 10^5 s + 12 \times 10^{10}}$$

$$= \frac{120 \times 10^{10}}{s^2 + 323076.9s + 12 \times 10^{10}}.$$

The loop gain of the current loop is

$$T_i(s) \approx \frac{12 f_s^2}{s \left(s + \frac{1 - a}{1 + a} 6 f_s\right)} = \frac{12 \times 10^{10}}{s(s + 323,076.9)}.$$

13.8 A buck-boost converter has $D = 0.407$, $V_O = -28 \, \text{V}$, $M_1 = 0.144 \, \text{A/s}$, $M_3 = 0.0986 \, \text{A/s}$, $\Delta V_I = 1 \, \text{V}$. Find K_i and ΔD. Assuming that the voltage loop is open and the feedforward voltage is the only change in the circuit, calculate the output voltage V_O.

The feedforward coefficient is

$$K_i = -\frac{D^2}{\left(1 + \frac{M_3}{M_1}\right)(1 - D)V_O} = -\frac{0.407^2}{\left(1 + \frac{0.0986}{0.144}\right)(1 - 0.407)(-28)} = 0.059.$$

The change in the duty cycle is

$$\Delta D = K_i \Delta V_I = 0.059 \times 1 = 0.059 = 5.9 \%.$$

The new value of the duty cycle is

$$D = D(0^-) + \Delta D = 0.407 + 0.059 = 0.466.$$

Hence, the dc output voltage will change from -28 V to

$$V_O = -\frac{D\eta}{1-D} V_I = -\frac{0.466 \times 0.85}{1 - 0.466} \times 42 = -31.154 \, \text{V}.$$

Chapter 14

Current-mode Control of Boost Converter

.1 The boost converter designed in Chapter 3 has $V_{Inom} = 156\,\text{V}$, $V_O = 400\,\text{V}$, $D_{nom} = 0.65$, $R_{Lmin} = 1.778\,\text{k}\Omega$, $r_{DS} = 1\,\Omega$, $V_F = 1.4\,\text{V}$, $R_F = 0.0171\,\Omega$, $L = 30\,\text{mH}$, $r_L = 2.1\,\Omega$, $C = 1\,\mu\text{F}$, and $r_C = 1\,\Omega$. Determine z_{i1} and f_{zi1}.

The LHP zero is

$$z_{i1} = -\frac{1}{C(R_{Lmin}/2 + r_C)} = -\frac{1}{10^{-6} \times (1778/2 + 1)} = -1124\,\text{rad/s},$$

and the frequency of the LHP zero is

$$f_{zi1} = -\frac{z_{i1}}{2\pi} = -\frac{-1124}{2\pi} = 178.8\,\text{Hz}.$$

.2 For the boost converter of Problem 14.1, determine T_{pio} and T_{pix}.

The duty cycle-to-inductor current transfer function at $f = 0$ is

$$T_{pio} = \frac{2V_O}{(1 - D_{nom})^2 R_{Lmin} + r} = \frac{2 \times 400}{(1 - 0.65)^2 \times 1778 + 2.756}$$
$$= 3.627\,\text{A} = 11.19\,\text{dBA},$$

and

$$T_{pix} = \frac{V_O(R_{Lmin} + 2r_C)}{L(R_{Lmin} + r_C)} = \frac{400 \times (1778 + 2 \times 1)}{30 \times 10^{-6}(1778 + 1)}$$
$$= 13.318 \times 10^6\,\text{A} = 142.5\,\text{dBA}.$$

14.3 For the boost converter of Problem 14.1, determine M_{vio} and f_{zi2}.

The input-to-output transfer function at $f = 0$ is

$$M_{vio} = \frac{1}{(1 - D_{nom})^2 R_{Lmin} + r} = \frac{1}{(1 - 0.65)^2 \times 1778 + 2.756}$$
$$= 0.0045 \, \text{A/V} = -46.78 \, \text{dBA/V},$$

and

$$f_{zi2} = \frac{1}{2\pi C (R_{Lmin} + r_C)} = \frac{1}{2 \times \pi \times 10^{-6} \times (1778 + 1)} = 89.46 \, \text{Hz}.$$

14.4 For the boost converter of Problem 14.1, determine A_{io} and f_{zn}.

The output current-to-inductor current transfer function at $f = 0$ is

$$A_{io} = \frac{(1 - D_{nom}) R_{Lmin}}{(1 - D_{nom})^2 R_{Lmin} + r} = \frac{(1 - 0.65) \times 1778}{(1 - 0.65)^2 \times 1778 + 2.756}$$
$$= 2.82 = 9 \, \text{dB}$$

and

$$f_{zn} = \frac{1}{2\pi C r_C} = \frac{1}{2 \times \pi \times 10^{-6} \times 1} = 159 \, \text{kHz}.$$

Chapter 15

Silicon and Silicon Carbide Power Diodes

.1 A silicon sample is doped with 1.5×10^{16} boron atoms/cm^3. Find the electron and hole concentrations and the semiconductor resistivity and conductivity at $T = 300$ K. What is the ratio of the silicon to boron atoms? What is the ratio p/n? What is the ratio of the resistivities of the intrinsic to doped silicon at $T = 300$ K?

Boron is an acceptor dopant. At $T = 300$ K, complete ionization occurs. Charge neutrality requires that the total positive and negative charge in an acceptor doped semiconductor is the same. Hence, the concentration of holes is

$$p = N_A + n \approx N_A = 1.5 \times 10^{16} \text{ holes/cm}^3.$$

The concentration of electrons is

$$n = \frac{n_i^2}{p} = \frac{p_i^2}{p} = \frac{p_i^2}{N_A} = \frac{(1.5 \times 10^{10})^2}{1.5 \times 10^{16}} = 1.5 \times 10^4 \text{ electrons/cm}^3.$$

There are 5×10^{22} silicon atoms/cm^3. The ratio of the silicon to boron atoms is

$$\frac{5 \times 10^{22}}{1.5 \times 10^{16}} = 3.333 \times 10^6.$$

Thus, the silicon is doped with one boron atom for every 333,333 silicon atoms. The ratio p/n is

$$\frac{p}{n} = \frac{1.5 \times 10^{16}}{1.5 \times 10^4} = 10^{12}.$$

The doped silicon resistivity is

$$\rho_p = \frac{1}{qN_A\mu_p} = \frac{1}{1.6 \times 10^{-19} \times 1.5 \times 10^{16} \times 480} = 0.868 \ \Omega \text{ cm}.$$

The doped silicon conductivity is

$$\sigma_p = qN_A\mu_p = \frac{1}{\rho_p} = \frac{1}{0.868} = 1.152\,(\Omega\,\text{cm})^{-1}.$$

The resistivity of the intrinsic silicon is

$$\rho_{Si} = \frac{1}{qn\mu_n + qp\mu_p} = \frac{1}{qn_i(\mu_n + \mu_p)} = \frac{1}{1.6 \times 10^{-19} \times 1.5 \times 10^{10}(1360 + 480)}$$

$$= 2.26 \times 10^5\,\Omega\,\text{cm},$$

and the intrinsic silicon conductivity is

$$\sigma_{Si} = qn\mu_n + qp\mu_p = \frac{1}{\rho_{Si}} = \frac{1}{2.26 \times 10^5} = 4.4424 \times 10^{-6}\,(\Omega\,\text{cm})^{-1}.$$

Hence, the ratio of the intrinsic to doped silicon resistivities is given by

$$\frac{\rho_{Si}}{\rho_p} = \frac{2.26 \times 10^5}{1.098} = 2.058 \times 10^5.$$

Thus, the semiconductor resistivity can be controlled by the addition of dopants into t
silicon lattice that contribute either holes (as in this case) or electrons to the conducti
process when ionized.

15.2 A silicon step junction diode has $N_D = 10^{16}\,\text{cm}^{-3}$, $N_A = 10^{18}\,\text{cm}^{-3}$, $T = 300\,\text{K}$, a
$V_D = 0$. Find x_p, x_n, W, and x_n/x_p.

The built-in potential is

$$V_{bi} = V_T \ln \frac{N_A N_D}{n_i^2} = 0.026 \ln \frac{10^{18} \times 10^{16}}{(1.5 \times 10^{10})^2} = 0.8171\,\text{V}.$$

The depletion width on the p-side is

$$x_p = \sqrt{\frac{2\epsilon_r \epsilon_0 V_{bi}}{qN_A\left(1 + \frac{N_A}{N_D}\right)}} = \sqrt{\frac{2 \times 11.7 \times 8.8542 \times 10^{-14} \times 0.8171}{1.6 \times 10^{-19} \times 10^{18}\left(1 + \frac{10^{18}}{10^{16}}\right)}} = 0.32\,\mu\text{m}.$$

The depletion width on the n-side is

$$x_n = \frac{N_A}{N_D}x_p = \frac{10^{18}}{10^{16}} \times 0.32 = 32\,\mu\text{m}.$$

The total depletion width is

$$W = x_p + x_n = 0.32 + 32 = 32.32\,\mu\text{m}.$$

The ratio is

$$\frac{x_n}{x_p} = \frac{32}{0.32} = 100.$$

.3 A silicon step junction diode has $N_D = 10^{14}\,\mathrm{cm^{-3}}$, $N_A = 10^{16}\,\mathrm{cm^{-3}}$, $T = 300\,\mathrm{K}$, and $V_D = 0$. Find x_p, x_n, W, and x_n/x_p.

The built-in potential is

$$V_{bi} = V_T \ln \frac{N_A N_D}{n_i^2} = 0.026 \ln \frac{10^{16} \times 10^{14}}{(1.5 \times 10^{10})^2} = 0.578\,\mathrm{V}.$$

The depletion width on the n-side is

$$x_n = \sqrt{\frac{2\epsilon_r \epsilon_0 V_{bi}}{q N_D \left(1 + \dfrac{N_D}{N_A}\right)}} = \sqrt{\frac{2 \times 11.7 \times 8.8542 \times 10^{-14} \times 0.578}{1.6 \times 10^{-19} \times 10^{20} \left(1 + \dfrac{10^{14}}{10^{16}}\right)}} = 0.272\,\mathrm{\mu m}.$$

The depletion width on the p-side is

$$x_p = \frac{N_D}{N_A} x_n = \frac{10^{14}}{10^{16}} \times 0.272 = 0.00272\,\mathrm{\mu m} = 2.72\,\mathrm{nm}.$$

The depletion width is

$$W = x_p + x_n = 0.272 + 0.00272 = 0.2747\,\mathrm{\mu m}.$$

The ratio is

$$\frac{x_n}{x_p} = \frac{0.272}{0.00272} = 100.$$

.4 A silicon step junction diode has $N_D = 10^{14}\,\mathrm{cm^{-3}}$, $N_A = 10^{16}\,\mathrm{cm^{-3}}$, $T = 300\,\mathrm{K}$, and $v_D = -600\,\mathrm{V}$. Find x_p, x_n, W, x_n/x_p, and E_m.

The built-in potential is

$$V_{bi} = V_T \ln \frac{N_A N_D}{n_i^2} = 0.026 \ln \frac{10^{16} \times 10^{14}}{(1.5 \times 10^{10})^2} = 0.578\,\mathrm{V}.$$

The depletion width on the n-side is

$$x_n = \sqrt{\frac{2\epsilon_r \epsilon_0 (V_{bi} - v_D)}{q N_D \left(1 + \dfrac{N_D}{N_A}\right)}} = \sqrt{\frac{2 \times 11.7 \times 8.8542 \times 10^{-14} \times (0.578 + 600)}{1.6 \times 10^{-19} \times 10^{14} \left(1 + \dfrac{10^{14}}{10^{16}}\right)}}$$

$$= 88\,\mathrm{\mu m}.$$

The depletion width on the p-side is

$$x_p = \frac{N_D}{N_A} x_n = \frac{10^{14}}{10^{16}} \times 88 = 0.88\,\mathrm{\mu m}.$$

The depletion width is

$$W = x_p + x_n = 0.88 + 88 = 88.88\,\mathrm{\mu m}.$$

The maximum electric field intensity is

$$E_m = \sqrt{\frac{2qN_D(V_{bi} - v_D)}{\epsilon_r \epsilon_0 \left(1 + \dfrac{N_D}{N_A}\right)}} = \sqrt{\frac{2 \times 1.6 \times 10^{-19} \times 10^{14}(0.578 + 600)}{11.7 \times 8.8542 \times 10^{-14}\left(1 + \dfrac{10^{14}}{10^{16}}\right)}}$$

$$= 135.5287 \, \text{kV/cm}.$$

15.5   At time $t = 0$, the current of a p^+n diode is increased from 0 to I_F with the rise time equal to zero. Derive an expression for the charge waveform $Q_p(t)$ and the diode voltage waveform $v_D(t)$. Draw these waveforms using MATLAB. Neglect the junction capacitance.

The excess minority hole charge equation is

$$\frac{dQ_p}{dt} = i_D - \frac{Q_p}{\tau_p}.$$

Since $i_D = I_F$, the charge equation for our situation becomes

$$\frac{dQ_p}{dt} = I_F - \frac{Q_p}{\tau_p}.$$

Hence,

$$dt = \frac{dQ_p}{I_F - \dfrac{Q_p}{\tau_p}},$$

resulting in

$$\int_0^t dt = \int_0^{Q_p(t)} \frac{dQ_p}{I_F - \dfrac{Q_p}{\tau_p}}.$$

Thus,

$$t = -\tau_p \ln\left(I_F - \frac{Q_p(t)}{\tau_p}\right)\bigg|_0^{Q_p(t)} = -\tau_p \ln\left[1 - \frac{Q_p(t)}{\tau_p I_F}\right].$$

Rearranging this equation, we obtain the excess minority hole charge waveform

$$Q_p(t) = \tau_p I_F \left(1 - e^{-\frac{t}{\tau_p}}\right).$$

The excess minority hole charge waveform increases exponentially from 0 to τ_p. Assuming a quasi-static situation, where $dQ_p/dt = 0$, we can write

$$Q_p(t) = \tau_p i_D = \tau_p I_S \left(e^{\frac{v_D}{nV_T}} - 1\right).$$

Equating both equations for $Q_p(t)$, we obtain the diode voltage waveform

$$v_D(t) = nV_T \ln\left[1 + \frac{I_F}{I_S}\left(1 - e^{-\frac{t}{\tau_p}}\right)\right].$$

5.6 At time $t = 0$, the current of a p$^+$n diode i_D is suddenly reduced from I_F to I_R with the fall time equal to zero. Derive an expression for the excess minority charge $Q_p(t)$ and draw it using MATLAB for $I_F = 1$ A, $I_R = -5$ A, and $\tau_p = 1200$ ns. Derive an expression for the storage time t_s. Calculate t_s in terms of τ_p for: (a) $I_R = -0.1 I_F$\$. (b) $I_R = -I_F$. (c) $I_R = -10 I_F$.

The excess minority hole charge equation is

$$\frac{dQ_p}{dt} = i_D - \frac{Q_p}{\tau_p}.$$

Because $i_D = I_R$, we obtain the charge equation for our case

$$\frac{dQ_p}{dt} = I_R - \frac{Q_p}{\tau_p},$$

which gives

$$dt = \frac{dQ_p}{I_R - \dfrac{Q_F}{\tau_p}}.$$

Next,

$$\int_0^{t_s} dt = \int_{Q_F(0)}^{Q_p(t)} \frac{dQ_p}{I_R - \dfrac{Q_p}{\tau_p}}.$$

The charge waveform can be obtained as follows:

$$t = -\tau_p \ln \left(\frac{Q_p}{\tau_p} - I_R \right) \Bigg|_{Q_F(0)}^{Q_p(t)} = -\tau_p \ln \left[\frac{I_R - \dfrac{Q_p(t)}{\tau_p}}{I_R - \dfrac{Q_F(0)}{\tau_p}} \right].$$

Rearrangement of this equation yields the excess stored charge

$$Q_p(t) = \tau_p I_R + \tau_p \left[\frac{Q_F(0)}{\tau_p} - I_R \right] e^{-\frac{t}{\tau_p}}.$$

If the diode current is constant and equal to I_F before the turn-off transition and the excess charge reached steady state after the turn-on transition,

$$\frac{Q_F(0)}{\tau_p} = I_F.$$

In this case, the expression for the charge waveform simplifies to

$$Q_p(t) = \tau_p I_R + \tau_p (I_F - I_R) e^{-\frac{t}{\tau_p}}.$$

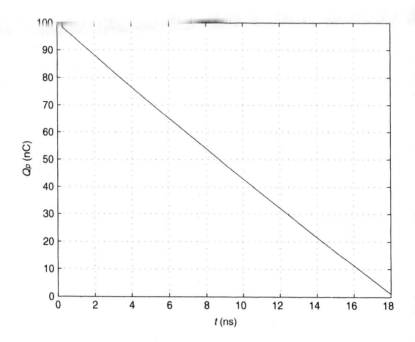

Since $Q_p(t_s) = 0$,

$$\int_0^{t_s} dt = \int_{Q_F(0)}^0 \frac{dQ_p}{I_R - \dfrac{Q_p}{\tau_p}}.$$

Because $Q_F(0) = \tau_p I_F$, we obtain a general expression for the storage time

$$t_s = \tau_p \ln \left(\frac{Q_p}{\tau_p} - I_R \right) \Bigg|_{Q_F(0)}^0 = \tau_p \ln \left[\frac{I_R - \dfrac{Q_F(0)}{\tau_p}}{I_R} \right] = \tau_p \ln \left(1 - \frac{I_F}{I_R} \right)$$

$$= \tau_p \ln \left(1 + \frac{I_F}{|I_R|} \right).$$

For $I_R = -0.1 I_F$, the storage time is

$$t_s = \tau_p \ln \left(1 + \frac{I_F}{|I_R|} \right) = \tau_p \ln \left(1 - \frac{1}{-0.1} \right) = 2.4 \tau_p.$$

For $I_R = -I_F$, the storage time is

$$t_s = \tau_p \ln \left(1 + \frac{I_F}{|I_R|} \right) = \tau_p \ln(1 + 1) = 0.69 \tau_p.$$

For $I_R = -10I_F$, the storage time is

$$t_s = \tau_p \ln\left(1 + \frac{I_F}{|I_R|}\right) = \tau_p \ln(1 + 0.1) = 0.0953\tau_p.$$

As $|I_R|/I_F$ increases, the storage time t_s decreases.

5.7 A dc current source with current I_F supplies a p^+n junction diode in the forward direction for a long time and at $t = 0$ is suddenly removed. Derive an expression for the excess minority hole charge and draw $Q_p(t)/(\tau_p I_F)$ as a function of t/τ_p using MATLAB. Find the time in terms of τ_p at which the initial charge decreases to 10% of its initial value.

The excess minority hole charge equation is

$$\frac{Q_p}{\tau_p} + \frac{dQ_p}{dt} = i_D(t) = 0.$$

Hence,

$$\frac{Q_p(s)}{\tau_s} + sQ_p(s) - Q_p(0) = 0,$$

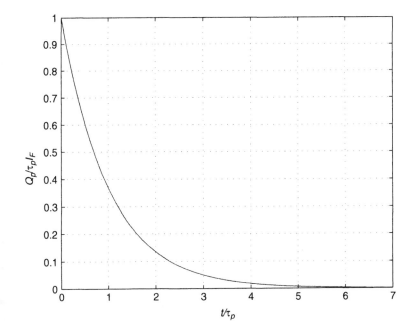

resulting in

$$Q_p(s) = \frac{Q_p(0)}{s + \dfrac{1}{\tau_p}}.$$

Thus,

$$Q_p(t) = Q_p(0)e^{-\frac{t}{\tau_p}}.$$

Since the diode was forward-biased with a dc current for a long time, it was in stead state, so that $Q_p(0) = \tau_p I_F$ and therefore

$$Q_p(t) = \tau_p I_F e^{-\frac{t}{\tau_p}}.$$

The excess minority hole charge dies out exponentially from the initial value $Q_p(0)$ $\tau_p I_F$ to zero with a time constant equal to the hole lifetime τ_p in the n-region. T excess charge $Q_p(t)$ approaches zero as time approaches infinity. The excess holes the n-region must die out by recombination only because $i_D = 0$ for $t > 0$.

To find the time when the excess charge decreases to $0.1Q_p(0)$, we write

$$0.1\tau_p I_F = \tau_p I_F e^{-\frac{t_{10\%}}{\tau_p}}$$

yielding

$$t_{10\%} = -\tau_p \ln 0.1 = 2.3\tau_p.$$

15.8   A silicon carbide pn junction diode has $di_F/dt < 0$, $f_s = 1\,\text{MHz}$, $V_R = -600\,\text{V}$, I_R $-5\,\text{A}$, and $t_t = 10\,\text{ns}$. Find the power dissipated during time t_t.

The power loss during time t_t is

$$P_{RR} = \frac{f_s V_R I_R t_t}{6} = \frac{10^6 \times (-600) \times (-5) \times (10 \times 10^{-9})}{6} = 5\,\text{W}.$$

15.9   Design a SiC Schottky diode to meet the following specifications: $V_{BD} = 600\,\text{V}$, I_{Dmax} $5\,\text{A}$, $J_{max} = 100\,\text{A/cm}^2$, and $V_{bi} = 10\,\text{V}$. Find: A_J, N_D, l_n, C_{J0}, and R_{DR}.

The diode cross-sectional area is

$$A_J = \frac{I_{Dmax}}{J_{max}} = \frac{5}{100} = 0.05\,\text{cm}^2 = 5 \times 10^{-6}\,\text{m}^2.$$

The minimum concentration is

$$N_{Dmin} = \frac{1.2974 \times 10^{19}}{V_{BD}} = \frac{1.2974 \times 10^{19}}{600} = 2.1623 \times 10^{16}\,\text{cm}^{-3}$$

$$= 2.1623 \times 10^{22}\,\text{m}^{-3}.$$

The minimum length of the n-region is

$$l_n = \frac{V_{BD}}{110} = \frac{600}{110} = 5.45\,\mu\text{m}.$$

Let $l_n = 6\,\mu\text{m}$. The zero-bias junction capacitance is

$$C_{J0} = A_J \sqrt{\frac{q\epsilon_r\epsilon_0 N_D}{2V_{bi}}}$$

$$= 5 \times 10^{-6} \sqrt{\frac{2 \times 1.602 \times 10^{-19} \times 9.7 \times 8.8542 \times 10^{-12} \times 2.1623 \times 10^{22}}{2 \times 10}}$$

$$= 864.2\,\text{pF}.$$

The drift region resistance is

$$R_{DR} = \rho\frac{l_n}{A_J} = \frac{l_n}{q\mu_n N_D A_J} = \frac{6 \times 10^{-6}}{1.602 \times 10^{-19} \times 900 \times 2.1623 \times 10^{22} \times 5 \times 10^{-6}}$$

$$= 0.3849\,\mu\Omega.$$

Chapter 16

Silicon and Silicon Carbide Power MOSFETs

6.1 The drift velocity of silicon electrons at $E = 100\,\text{kV/cm}$ is $10^7\,\text{cm/s}$. Find μ_n.

The mobility of silicon electrons is

$$\mu_n = \frac{v_n}{E} = \frac{10^7}{10^5} = 100\,\text{cm}^2/\text{Vs}.$$

6.2 Calculate the minimum thickness of the drift region for the power MOSFET with the breakdown voltage $V_{BD} = 50\,\text{V}$, $100\,\text{V}$, $200\,\text{V}$, and $400\,\text{V}$.

For $V_{BD} = 50\,\text{V}$,

$$W_D = \frac{V_{BD}}{10} = \frac{50}{10} = 5\,\mu\text{m}.$$

For $V_{BD} = 100\,\text{V}$,

$$W_D = \frac{V_{BD}}{10} = \frac{100}{10} = 10\,\mu\text{m}.$$

For $V_{BD} = 200\,\text{V}$,

$$W_D = \frac{V_{BD}}{10} = \frac{200}{10} = 20\,\mu\text{m}.$$

For $V_{BD} = 400\,\text{V}$,

$$W_D = \frac{V_{BD}}{10} = \frac{400}{10} = 40\,\mu\text{m}.$$

6.3 Calculate the maximum doping density of the drift region for the n-channel power MOSFET with the breakdown voltage $V_{BD} = 50\,\text{V}$, $100\,\text{V}$, $200\,\text{V}$, and $400\,\text{V}$.

For $V_{BD} = 50\,\text{V}$,

$$N_D = \frac{1.293 \times 10^{17}}{V_{BD}} = \frac{1.293 \times 10^{17}}{50} = 2.586 \times 10^{15}\,\text{cm}^{-3}.$$

For $V_{BD} = 100\,\text{V}$,

$$N_D = \frac{1.293 \times 10^{17}}{V_{BD}} = \frac{1.293 \times 10^{17}}{100} = 1.293 \times 10^{15}\,\text{cm}^{-3}.$$

For $V_{BD} = 200\,\text{V}$,

$$N_D = \frac{1.293 \times 10^{17}}{V_{BD}} = \frac{1.293 \times 10^{17}}{200} = 6.465 \times 10^{14}\,\text{cm}^{-3}.$$

For $V_{BD} = 400\,\text{V}$,

$$N_D = \frac{1.293 \times 10^{17}}{V_{BD}} = \frac{1.293 \times 10^{17}}{400} = 3.232 \times 10^{14}\,\text{cm}^{-3}.$$

The doping concentration N_D of the drift region must be reduced to achieve a high breakdown voltage V_{BD}.

16.4   The donor concentration in the silicon n-channel power MOSFET is (a) $N_D = 10^{14}\,\text{cm}^{-3}$ (b) $N_D = 10^{15}\,\text{cm}^{-3}$, and (c) $N_D = 10^{16}\,\text{cm}^{-3}$. Find the breakdown voltage.

(a) For $N_D = 10^{14}\,\text{cm}^{-3}$,

$$V_{BD} = \frac{\epsilon_{r(Si)}\epsilon_0 E_{BD}^2}{2qN_D} = \frac{11.7 \times 8.8542 \times 10^{-14} \times (2 \times 10^5)^2}{2 \times 1.6 \times 10^{-19} \times 10^{14}} = 1295\,\text{V}.$$

(b) For $N_D = 10^{15}\,\text{cm}^{-3}$,

$$V_{BD} = \frac{\epsilon_{r(Si)}\epsilon_0 E_{BD}^2}{2qN_D} = \frac{11.7 \times 8.8542 \times 10^{-14} \times (2 \times 10^5)^2}{2 \times 1.6 \times 10^{-19} \times 10^{15}} = 129.5\,\text{V}.$$

(c) For $N_D = 10^{16}\,\text{cm}^{-3}$,

$$V_{BD} = \frac{\epsilon_{r(Si)}\epsilon_0 E_{BD}^2}{2qN_D} = \frac{11.7 \times 8.8542 \times 10^{-14} \times (2 \times 10^5)^2}{2 \times 1.6 \times 10^{-19} \times 10^{16}} = 12.95\,\text{V}.$$

16.5   The donor concentration in the silicon-carbide n-channel power MOSFET is (a) N_D $10^{14}\,\text{cm}^{-3}$, (b) $N_D = 10^{15}\,\text{cm}^{-3}$, and (c) $N_D = 10^{16}\,\text{cm}^{-3}$. Find the breakdown voltag

(a) For $N_D = 10^{14}\,\text{cm}^{-3}$,

$$V_{BD} = \frac{\epsilon_{r(SiC)}\epsilon_0 E_{BD}^2}{2qN_D} = \frac{9.7 \times 8.8542 \times 10^{-14} \times (2.2 \times 10^6)^2}{2 \times 1.6 \times 10^{-19} \times 10^{14}} = 129.9\,\text{kV}.$$

(b) For $N_D = 10^{15}\,\text{cm}^{-3}$,

$$V_{BD} = \frac{\epsilon_{r(SiC)}\epsilon_0 E_{BD}^2}{2qN_D} = \frac{9.7 \times 8.8542 \times 10^{-14} \times (2.2 \times 10^6)^2}{2 \times 1.6 \times 10^{-19} \times 10^{15}} = 12.99\,\text{kV}.$$

(c) For $N_D = 10^{16}\,\text{cm}^{-3}$,

$$V_{BD} = \frac{\epsilon_{r(SiC)}\epsilon_0 E_{BD}^2}{2qN_D} = \frac{9.7 \times 8.8542 \times 10^{-14} \times (2.2 \times 10^6)^2}{2 \times 1.6 \times 10^{-19} \times 10^{16}} = 1.299\,\text{kV}.$$

6.6 A silicon MOSFET channel has $\mu_n = 600\,\text{cm}^2/\text{Vs}$. Calculate C_{ox} and k_p for: (a) $t_{ox} = 1\,\mu\text{m}$, (b) $t_{ox} = 0.1\,\mu\text{m}$, and (c) $t_{ox} = 0.01\,\mu\text{m}$.

(a) For $t_{ox} = 1\,\mu\text{m}$,

$$C_{ox} = \frac{\epsilon_{r(SiO_2)}\epsilon_0}{t_{ox}} = \frac{3.9 \times 8.8542 \times 10^{-12}}{1 \times 10^{-6}}$$

$$= 34.53\,\mu\text{F}/\text{m}^2 = 3.453\,\text{nF}/\text{cm}^2 = 34.53\,\text{pF}/\text{mm}^2$$

and

$$k_p = \mu_n C_{ox} = 0.06 \times 34.53 \times 10^{-6}\,\text{A}/\text{V}^2 = 2.07\,\mu\text{A}/\text{V}^2.$$

(b) For $t_{ox} = 0.1\,\mu\text{m}$,

$$C_{ox} = \frac{\epsilon_{r(SiO_2)}\epsilon_0}{t_{ox}} = \frac{3.9 \times 8.8542 \times 10^{-12}}{0.1 \times 10^{-6}}$$

$$= 345.3\,\mu\text{F}/\text{m}^2 = 34.53\,\text{pF}/\text{cm}^2 = 345.3\,\text{pF}/\text{mm}^2$$

and

$$k_p = \mu_n C_{ox} = 0.06 \times 345.3 \times 10^{-6}\,\text{A}/\text{V}^2 = 20.7\,\mu\text{A}/\text{V}^2.$$

(c) For $t_{ox} = 0.01\,\mu\text{m}$,

$$C_{ox} = \frac{\epsilon_{r(SiO_2)}\epsilon_0}{t_{ox}} = \frac{3.9 \times 8.8542 \times 10^{-12}}{0.01 \times 10^{-6}}$$

$$= 3453\,\mu\text{F}/\text{m}^2 = 345.3\,\text{nF}/\text{cm}^2 = 3.453\,\text{nF}/\text{mm}^2$$

and

$$k_p = \mu_n C_{ox} = 0.06 \times 3453 \times 10^{-6}\,\text{A}/\text{V}^2 = 207\,\mu\text{A}/\text{V}^2.$$

6.7 A power MOSFET has $\mu_n C_{ox} = 20\,\mu\text{A}/\text{V}^2$, $W/L = 10^4$, $V_t = 3\,\text{V}$, $\theta = 0.1$, and $V_{GS} = 10\,\text{V}$. Find the drain saturation current at the boundary between the linear and the saturation region I_{Dsat} for a long-channel MOSFET, the long-channel resistance R_c, the short-channel resistance $R_{c(sc)}$, and the ratio $R_{c(sc)}/R_c$.

The drain current at the boundary between the linear and the saturation regions I_{Dsat} for a long channel is

$$I_{Dsat} = \frac{1}{2}\mu_n C_{ox}\left(\frac{W}{L}\right)(V_{GS} - V_t)^2 = \frac{1}{2} \times 20 \times 10^{-6} \times 10^4(10 - 3)^2 = 4.9\,\text{A}.$$

The channel resistance R_c for a long channel is

$$R_c = \frac{1}{\mu_n C_{ox}} \left(\frac{L}{W} \right) \frac{1}{V_{GS} - V_t} = \frac{1}{20 \times 10^{-6}} \times 10^{-4} \frac{1}{10 - 3} = 0.714\,\Omega.$$

The channel resistance $R_{c(sc)}$ for a short channel is

$$R_{c(sc)} = \frac{1}{\mu_n C_{ox}} \left(\frac{L}{W} \right) \left(\theta + \frac{1}{V_{GS} - V_t} \right) = \frac{1}{20 \times 10^{-6}} \times 10^{-4} \left(0.1 + \frac{1}{10 - 3} \right)$$

$$= 1.214\,\Omega.$$

The ratio is

$$\frac{R_{c(sc)}}{R_c} = \frac{1.214}{0.714} = 1.7.$$

16.8 A power MOSFET has $L = 1\,\mu m$, $W/L = 1.6 \times 10^6$, $W_p = 4\,\mu m$, and $d_{pp} = 2\,\mu m$. Find the overall channel width W, the number of cells n, the area of all cells A_c, the cell density CD, the channel width per unit area D, and the chip area utilization U. Assume that the source pads, gate pads, and the field guard rings increase the chip area by 25%.

The total channel width W is given by

$$W = (W/L)L = 1.6 \times 10^6 \times 1 \times 10^{-6} = 1.6\,m = 1.6 \times 10^6\,\mu m.$$

The number of cells is

$$n = \frac{W}{4W_p} = \frac{1.6 \times 10^6}{4 \times 4} = 10^5.$$

The area of a single cell is

$$A_1 = (W_p + d_{pp})^2 = (4 \times 10^{-6} + 2 \times 10^{-6})^2 = 36 \times 10^{-12}\,m^2 = 36\,\mu m^2.$$

The area of all cells

$$A_c = nA_1 = 10^5 \times 36 \times 10^{-12} = 3.6 \times 10^{-6}\,m^2 = 3.6\,mm^2.$$

The cell density is

$$CD = \frac{n}{A_c} = \frac{10^5}{3.6 \times 10^{-6}} = 27,778 \times 10^6\,cells/m^2 = 27,778\,cells/mm^2.$$

The channel width per unit area is

$$D = \frac{W}{A_c} = \frac{1.6\,m}{3.6\,mm^2} = 0.4444 \times 10^6\,m/m^2 = 0.4444\,m/mm^2.$$

The chip area utilization is

$$U = \frac{W}{A_{ch}} = \frac{W}{1.25A_c} = \frac{1.6}{1.25 \times 3.6 \times 10^{-6}} = 0.3556 \times 10^6\,m/m^2 = 0.3556\,m/mm^2.$$

6.9 Derive an expression for the channel width per unit area D for a power HEXFET and compare it to that of the power MOSFET with square cells.

The area of a regular polygon is

$$A = Nr^2 \tan\left(\frac{180°}{N}\right),$$

and the perimeter of a regular polygon is

$$W = 2Nr\tan\left(\frac{180°}{N}\right),$$

where r is the radius of the inscribed circle and N is the number of sides.

For the power MOSFET cell, the inscribed radius of the entire cell $r_1 = (W_p + d_{pp})/2$, and the inscribed radius of the cell encircled by the channel $r_2 = W_p/2$. Thus, the channel width per unit area for any polygon is

$$\frac{W}{A_c} = \frac{W_1}{A_1} = \frac{4W_p}{(W_p + d_{pp})^2}.$$

For the HEXFET, $N = 6$, the inscribed radius of the entire cell $r_1 = (W_p + d_{pp})/2$, and the inscribed radius of the cell encircled by the channel $r_2 = W_p/2$. Hence, we obtain the area of a single cell

$$A_1 = Nr_1^2 \tan\left(\frac{180°}{N}\right) = 6\left(\frac{W_p + d_{pp}}{2}\right)^2 \tan\left(\frac{180°}{6}\right) = 6\left(\frac{W_p + d_{pp}}{2}\right)^2 \tan(30°)$$

$$= \frac{3}{2}(W_p + d_{pp})^2 \frac{1}{\sqrt{3}} = \frac{\sqrt{3}}{2}(W_p + d_{pp})^2,$$

and the area of n cells

$$A_c = nA_1 = \frac{\sqrt{3}}{2}n(W_p + d_{pp})^2.$$

The channel length of a single cell is

$$W_1 = 2Nr_2\tan\left(\frac{180°}{N}\right) = 2 \times 6\left(\frac{W_p}{2}\right)\tan\left(\frac{180°}{6}\right) = 6W_p\frac{1}{\sqrt{3}} = 2\sqrt{3}W_p,$$

and the channel width of n cells is

$$W = nW_1 = 2\sqrt{3}nW_p.$$

Hence, the channel width per unit area for the HEXFET is

$$D = \frac{W}{A_c} = \frac{2\sqrt{3}nW_p}{\frac{\sqrt{3}}{2}n(W_p + d_{pp})^2} = \frac{4W_p}{(W_p + d_{pp})^2}.$$

Thus, the channel width per unit area D is identical for the HEXFET and the power MOSFET with square cells.

16.10 Design a SiC n-channel power MOSFET for switching applications to meet the flowing specifications: $V_{DSS} = 1\,kV$, $I_{SM\,max} = 10\,A$, $r_{DS} \le 0.2\,\Omega$, $\mu_n = 400\,cm^2/Vs$, $V_t = 2\,V$ and $V_{GS\,max} = 10\,V$.

Assuming $t_{ox} = 0.1\,\mu m$,

$$C_{ox} = \frac{\epsilon_{r(SiO_2)}\epsilon_0}{t_{ox}} = \frac{3.9 \times 8.8542 \times 10^{-12}}{0.1 \times 10^{-6}} = 345.3\,\mu F/m^2 = 34.53\,nF/cm^2.$$

Hence,

$$k_p = \mu_n C_{ox} = 0.04 \times 345.3 \times 10^{-6} = 13.812\,\mu A/V^2.$$

Assuming $a = 0.05$, the minimum aspect ratio is

$$\left(\frac{W}{L}\right)_{min} = \frac{I_{SM\,max}}{\frac{1}{2}ak_p(V_{GS} - V_t)^2} = \frac{10}{\frac{1}{2} \times 0.05 \times 13.812 \times 10^{-6}(10 - 2)^2} = 4.525 \times 10$$

Let $L = 1\,\mu m$. Hence,

$$W = \left(\frac{W}{L}\right)_{min} L = 4.525 \times 10^5 \times 10^{-6} = 0.4525\,m.$$

Let $W_p = 20\,\mu m$, $d_{pp} = 10\,\mu m$, and $t_m = 1\,\mu m$. Thus,

$$W_1 = 4W_p = 4 \times 20 = 80\,\mu m$$

and

$$n = \frac{W}{W_1} = \frac{0.4525}{80 \times 10^{-6}} = 5.656 \times 10^3.$$

The area of all cells is

$$A_c = n(W_p + d_{pp})^2 = [(20 + 10) \times 10^{-6}]^2 \times 5.653 \times 10^3$$

$$= 5.0877 \times 10^{-6}\,m^2 = 5.0877\,mm^2.$$

The estimated die area is

$$A_{die} = 1.25A_c = 1.25 \times 5.0877 = 6.359625\,mm^2.$$

The maximum doping concentration is

$$N_{Dmax(SiC)} = \frac{1.2974 \times 10^{19}}{V_{BD(SiC)}} = \frac{1.2974 \times 10^{19}}{10^3} = 1.2974 \times 10^{16}\,donors/cm^3.$$

Pick $N_D = 10^{16}$ donors/cm^3. The minimum thickness of the drift region is

$$W_D = \frac{V_{BD}}{110} = \frac{1000}{110} = 9.09\,\mu m.$$

Pick $W_D = 12\,\mu m$.

The long-channel resistance is given by

$$R_{Ch} = \frac{1}{k_p\left(\dfrac{W}{L}\right)(V_{GS} - V_t)} = \frac{1}{13.812 \times 10^{-6} \times 4.525 \times 10^5 \times (10 - 2)} = 20\,m\Omega.$$

The accumulation region resistance is

$$R_a = \frac{1}{4k_p \left(\dfrac{W}{d_{pp}}\right)(V_{GS} - V_t)} = \frac{1}{4 \times 13.812 \times 10^{-6} \times (0.4525 \times 10^6/10) \times (10 - 2)}$$

$$= 50 \, \text{m}\Omega.$$

Assuming $h_p = 3 \, \mu\text{m}$, the neck region resistance is

$$R_n = \frac{h_p}{q\mu_n N_D (W d_{pp} + n d_{pp}^2)}$$

$$= \frac{3 \times 10^{-6}}{1.602 \times 10^{-19} \times 0.04 \times 10^{22} \times [(1 \times 10 \times 10^{-6}/2 + 5.653 \times 10^3 \times (10 \times 10^{-6})^2]}$$

$$= 8.347 \, \text{m}\Omega.$$

The drift region resistance is

$$R_{DR} = \frac{W_D}{q\mu_n N_D A_c} = \frac{12 \times 10^{-6}}{1.602 \times 10^{-19} \times 0.04 \times 10^{22} \times 5.0877 \times 10^{-6}} = 36.808 \, \text{m}\Omega.$$

The MOSFET on-resistance is

$$r_{DS} = R_{Ch} + R_a + R_n + R_{DR} = 20 + 50 + 8.347 + 36.808 = 115.14 \, \text{m}\Omega.$$

Chapter 17
Soft-switching DC–DC Converters

.1 Design a ZVS quasi-resonant converter to meet the following specifications: $V_O = 28\,\text{V}$, $V_I = 42\,\text{V}$, and $I_O = 2\,\text{A}$.

We will design a buck ZVS quasi-resonant half-wave converter. The output power is

$$P_O = I_O V_O = 2 \times 28 = 56\,\text{W}.$$

The load resistance is

$$R_L = \frac{V_O}{I_O} = \frac{28}{2} = 14\,\Omega.$$

The dc voltage transfer function is

$$M_{V\,DC} = \frac{V_O}{V_I} = \frac{28}{42} = 0.667.$$

Let us use the buck half-wave ZVS converter with $n = 1$, $Q = M_{V\,DC} = 0.667$, and $f_s = 1\,\text{MHz}$. The characteristic impedance of the resonant circuit is

$$Z_o = \frac{R_L}{Q} = \frac{14}{0.667} = 21\,\Omega.$$

The switching frequency is

$$f_0 = \frac{0.9092 f_s}{1 - M_{V\,DC}} = \frac{0.9092 \times 10^6}{1 - 0.667} = 2.73\,\text{MHz}.$$

The duty cycle is

$$D = 1 - 0.9092 \left(\frac{f_s}{f_0} \right) = 1 - 0.9092 \left(\frac{1}{2.73} \right) = 0.667.$$

The resonant inductance is

$$L_r = \frac{R_L}{\omega_0 Q} = \frac{14}{2\pi \times 2.73 \times 10^6 \times 0.667} = 1.224\,\mu\text{H}$$

and the resonant capacitance is

$$C_r = \frac{Q}{\omega_0 R_L} = \frac{0.667}{2\pi \times 2.73 \times 10^6 \times 14} = 2.7775 \, \text{nF}.$$

Pick $L_r = 1.2 \, \mu\text{H}$ and $C_r = 2.7 \, \text{nF}$.

The peak switch current is

$$I_{SM} = I_O = 2 \, \text{A}.$$

The peak switch voltage is

$$V_{SM} = \left(\frac{M_{V \, DC}}{Q} + 1 \right) V_I = 2V_I = 2 \times 42 = 84 \, \text{V}.$$

The peak diode current is

$$I_{DM} = 2I_O = 2 \times 2 = 4 \, \text{A}.$$

The peak diode voltage is

$$V_{DM} = V_I = 42 \, \text{V}.$$

17.4 Design a ZCS quasi-resonant converter to meet the following specifications: $V_O = 12$ $V_I = 24 \, \text{V}$, and $I_O = 5 \, \text{A}$.

A boost ZCS quasi-resonant full-wave converter will be designed. The output power

$$P_O = I_O V_O = 5 \times 24 = 120 \, \text{W}.$$

The load resistance is

$$R_L = \frac{V_O}{I_O} = \frac{24}{5} = 4.8 \, \Omega.$$

The dc voltage transfer function is

$$M_{V \, DC} = \frac{V_O}{V_I} = \frac{24}{12} = 2.$$

Let us use the buck half-wave ZVS converter with $Q = M_{V \, DC} = 2$, $n = 1$, and f_s 1 MHz. The characteristic impedance of the resonant circuit is

$$Z_o = \frac{R_L}{Q} = \frac{4.8}{2} = 2.4 \, \Omega.$$

The switching frequency is

$$f_0 = \frac{0.9092 f_s}{1 - \dfrac{1}{M_{V \, DC}}} = \frac{0.9092 \times 10^6}{1 - \dfrac{1}{2}} = 1.8184 \, \text{MHz}.$$

The duty cycle is

$$D = 0.9092 \left(\frac{f_s}{f_0} \right) = 0.9092 \frac{1}{1.8184} = 0.5.$$

The resonant inductance is

$$L_r = \frac{R_L}{\omega_0 Q} = \frac{4.8}{2\pi \times 1.8184 \times 10^6 \times 2} = 0.21\,\mu H$$

and the resonant capacitance is

$$C_r = \frac{Q}{\omega_0 R_L} = \frac{2}{2\pi \times 1.8184 \times 10^6 \times 4.8} = 36.4686\,nF.$$

Pick $L_r = 0.22\,\mu H$ and $C_r = 33\,nF$.

The peak switch current is

$$I_{SM} = (Q + M_{V\,DC})I_O = (2 + 2) \times 5 = 20\,A.$$

The peak switch voltage is

$$V_{SM} = V_O = 24\,V.$$

The peak diode current is

$$I_{DM} = I_I = M_{V\,DC}I_O = 2 \times 5 = 10\,A.$$

The peak diode voltage is

$$V_{DM} = 2V_O = 2 \times 24 = 48\,V.$$

Printed in the United States
By Bookmasters